异步编程实战

基于AIO的全栈开发技术

陈少佳◎编著

Chen Shaojia

清華大學出版社

北京

内容简介

本书系统地讲解了如何使用 Python 异步 IO 编程技术。从学习基础知识开始，展开讲解全栈框架的实现过程及项目开发过程中的实用技术。

本书分为基础篇和实战篇。基础篇(第1～7章)讲解 Python 异步 IO 的基础用法及常用技术框架的用法，包括 Python asyncio 库、页面渲染、异步 IO 数据库使用、ASGI 等技术；实战篇(第8章和第9章)介绍如何基于 Python 异步 IO 实现一个全栈框架，并利用该框架开发一个实用项目。本书配套源代码及视频教程，可以使读者学习起来更轻松。

本书适合有一定基础的开发人员阅读，用于突破职业瓶颈、从编码员进阶成架构师，也可作为培训机构的参考用书。

图书在版编目(CIP)数据

Python 异步编程实战：基于 AIO 的全栈开发技术/陈少佳编著. —北京：清华大学出版社，2021.3
(清华开发者书库 · Python)
ISBN 978-7-302-57683-9

Ⅰ. ①P… Ⅱ. ①陈… Ⅲ. ①软件工具－程序设计 Ⅳ. ①TP311.561

中国版本图书馆 CIP 数据核字(2021)第 045424 号

责任编辑：赵佳霓
封面设计：刘 键
责任校对：郝美丽
责任印制：宋 林

出版发行：清华大学出版社
网 址：http://www.tup.com.cn，http://www.wqbook.com
地 址：北京清华大学学研大厦 A 座 邮 编：100084
社 总 机：010-62770175 邮 购：010-83470235
投稿与读者服务：010-62776969，c-service@tup.tsinghua.edu.cn
质量反馈：010-62772015，zhiliang@tup.tsinghua.edu.cn
课件下载：http://www.tup.com.cn，010-83470236
印 装 者：大厂回族自治县彩虹印刷有限公司
经 销：全国新华书店
开 本：185mm×260mm **印 张**：20.25 **字 数**：493 千字
版 次：2021 年 5 月第 1 版 **印 次**：2021 年 5 月第1 次印刷
印 数：1～2000
定 价：79.00 元

产品编号：088578-01

序　一

FOREWORD

Python 作为一门优秀的编程语言，在各领域的出色表现，以及近些年的火爆不必多说，是有目共睹的！

异步编程在基于 C/C++ 语言进行系统编程为主流的年代虽然已经出现，但其编码结构，尤其是性能不尽人意，并且在 BSD/Linux/Windows 等操作系统采用的异步编程模型及库不尽相同，即使在相同的操作系统（例如各种 Linux 发行版）也经历了若干异步编程模型及库的升级，这就给软件架构增加了难度，并且有安全性和稳定性的方面隐患。

Python 在对异步编程模型方面的功能日益丰富和强大，这是个令人兴奋的进步。这使得过去许多复杂的架构问题变得简单易行，并且基于 Python 语法使得项目代码简洁高效，给代码的维护和升级，尤其是项目迭代带来了极大的便利。

全书从一些简单的案例和必要的基础知识开始，由浅入深、循序渐进地讲解了 Python 异步编程模型，并扩展了关联知识和实际应用案例，极大增加了本书的实用性，使读者既可以学习到相关知识，也可以参照本书中的案例完成自己的项目。

本书的作者陈少佳是我相识多年的哥们儿，既做过同事，也合作过项目。少佳的技术和人品在业界是有口皆碑的！在技术方面，从早期的桌面应用开发、前端开发、服务器端开发，到后来流行的手机 App 开发、游戏开发，再到现在的深度学习等技术，少佳始终走在技术前沿，平时除了项目开发，还会为不同类型的企业进行技术架构等方面的支持；在人品方面，无论在项目中，还是在生活中，少佳历来是言必行、行必果，在这个略显浮躁的年代，这是难得可贵的品质！

最后，也很感谢少佳邀请我对这样一本优秀的书籍作序，这是我作为技术人员出身的荣幸！相信各位读者会从本书中获益良多！

马锋

河南省行知教育研究院院长

国家开放大学软件学院教学处处长

中国软件行业协会教育与培训委员会特聘专家

2020 年 12 月 25 日

序　二

FOREWORD

编程是一种认识世界的思维方式。

世间万物包含了巨量的碎片化信息，而人类在对其组织形式、结构体系进行抽象、解构、归纳、总结后，形成了我们现有的门类众多的知识体系。而梳理事物的逻辑谱系也就是编写程序的思维基础了，可以说每一次的程序编写都是一次对人类知识的重新认知和阐释。在编程的过程中，程序设计师会反复地进行“发现问题——解决问题——发现问题”的循环，并逐步演化出独特的认识论和方法论。所以编程是一种非常有效的逻辑思维训练方式。

我们的教育体系一直是文理分科的培养模式，这种模式在特定历史时期的确高效地培养出了一批社会急需的人才，满足了工业化生产的需要。但是在当下的信息时代里，这种培养模式显得越来越无法满足社会的需要——理科生不懂艺术和美学，文科生则对自然科学一窍不通。

从功利性的角度来说，在某一个学科领域中被视为家常便饭的事情，可能对另一个学科而言就是至关重要的解决方案。例如3M公司有一款“手术薄膜”的产品，在进行手术时，病人的创口非常容易受到感染，覆盖上这款薄膜后再开刀就能有效降低感染。在进行薄膜的技术创新时，它们并不是从材料学专家那里找到解决方案，而是从兽医和好莱坞化妆师的工作中得到灵感——兽医的手术对象相比人类而言有更多的毛发，因此手术环境也更复杂并且创口也更容易感染，而化妆师则需要让化妆材料不会太过刺激到演员的皮肤，并且容易卸妆。最后3M公司成功开发出了一款抗感染的药膏，该款药膏的生产成本比薄膜还要低得多，深受发展中国家的医生和患者的欢迎。

从人的自我实现的角度来说，要成为一个完整的人，必须对美学艺术和科学技术都有所认知，不可有所偏废。例如爱因斯坦从小就培养出了对音乐的浓厚兴趣，并能够与专业的乐师一起进行演奏。艺术可以拓展想象力的边界，而科学则为幻想加入了燃料，二者一同推动着人类文明不断向前。

而编程则有助于培养自己用逻辑化的理性思维去看待艺术和科学这两个看似截然不同的学科。

初识本书作者陈少佳是在10年前的一个冬夜，我们对于游戏开发的想法和理念畅谈了许久。虽然那一次项目没有推进下去，但是让我和少佳成为好朋友。少佳的学识是最令我折服的——他在编程教育方面踏实耕耘了很多年，桃李遍天下；而我和少佳闲聊时也更愿意成为他的听众，聆听他对历史文化、神怪传说、自然科学等诸多方面的独到见解。

少佳精心编著的这本书是他多年编程生涯的凝练，其中理论结合实操，包含了诸多操作性非常强的编程案例，相信能够为各位读者的编程生涯插上进阶的翅膀！

王巍宾

北京天启创世科技有限公司 CEO

2020 年 11 月 17 日

前 言

PREFACE

Python 是一门功能强大的编程语言，业务领域非常广阔，涉及系统脚本、嵌入式开发、网站开发、游戏实时服务器、人工智能、大数据等，同时 Python 拥有简洁易懂的语法，学习难度不高，很适合作为学习编程的第一门语言。

最近几年异步编程模型兴起，大部分常用的编程语言都在向异步编程模型这个方向演化，甚至系统级编程语言 C++在 C++ 2020 的标准里也纳入了异步模型。当然 Python 也紧跟时代步伐支持了异步编程模型，在 Python 网站开发这个技术分支里已经涌现出了一系列的基于异步编程模型的框架，而一些老牌技术框架也开始逐步地支持异步编程，例如大名鼎鼎的 Django 在 3.0 版本以后支持了 ASGI，这只是个开始，相信在随后的发展中它会全面支持异步编程。

从服务器利用率层面来讲，采用异步编程模型可以有效利用服务器的 IO 资源，将服务器的硬件能力发挥到最大程度，从而节省运维成本。从开发层面来讲，异步编程模型可以用更加清晰整洁的代码来表达异步逻辑，从而节省开发成本。

所以现在到了我们必须学习 Python 异步编程的时候了。

在工作中，一个普通程序员容易遇到职业瓶颈，能力得不到提升、薪资无法增长，其根本原因是只会用框架而不懂框架的实现原理，当网站发展得越来越大时会出现一些问题，如果普通程序员无法解决这些问题，当然不会有晋升的机会，所以本书在讲解框架时争取做到进得去出得来，在基础篇（第 1～7 章）里讲解 Python 异步 IO 的基础知识和常用 Web 框架，深入 Web 开发的细节，之后跳出来纵观全局，于实战篇（第 8～9 章）里讲解如何基于异步 IO 实现一个完整的全栈 Web 框架，并以一个完整的实战项目来融会贯通所讲过的知识。

为了更有利于读者学习与实践，笔者将尽最大可能保证本书中的每个代码片断可独立运行。完整代码可扫描下方二维码下载。

本书源代码下载

希望本书能够对读者学习 Python 异步 IO 编程技术有所帮助，并恳请读者批评指正。

陈少佳

2020 年 12 月

目　录

CONTENTS

基　础　篇

实 战 篇

基础篇

基础篇包括第1～7章，讲解Python语言与asyncio库相关的技术，以及基于该技术发展而来的第三方库的使用方式。

第1章 Python AIO库

介绍Python内置的用于支持AIO编程的库，从基本概念及用法入手，以简单易懂的方式讲解与Python AIO编程相关的知识，并基于这些知识实现本机文件操作异步化、异步Socket及异步HTTP。

第2章 Docker工具

Docker是一个强大的容器化工具，可将大量的重复性操作抽象出来，使这些操作可以流程化、配置化，同样的流程和配置可以很方便地迁移到另一台机器上，从而大大提高开发与维护的效率。

本章将详细介绍Docker与Docker Compose工具的用法。

第3章 AIOHTTP

AIOHTTP是基于Python AIO实现的一套完整的Python Web开发框架，简单易用，社区活跃，在使用该框架开发网站时如果遇到框架本身的问题，可以在其GitHub仓库提交问题，总能得到及时回复与问题修复。

本章将详细讲解AIOHTTP的用法，包括路由机制、静态文件处理机制、模板渲染机制、表单提交与处理、Session支持等。

第4章 aioMySQL

本章介绍aioMySQL的用法，aioMySQL是由AIOHTTP开发团队开发的，将Python版MySQL驱动异步化，从而适应AIO编程，以便在由AIO编程技术构建的网站中完美使用MySQL数据库。

第5章 ASGI

ASGI是新一代通用网关接口技术规范，以全新的方式去适应AIO编程，本章将详细介绍与ASGI相关的技术。

第6章 Tornado

Tornado是一个经得起实践检验的、使用广泛的异步IO库，早在Python官方的

asyncio 之前就已经出现，虽然没有 asyncio 那么高效、华丽而简洁的语法，但是在 BIO 编程时代，Tornado 可谓一枝独秀，在旧的 Python 语法基础上提供了完整的异步 IO 编程解决方案。

在 Python 官方的 asyncio 出现之后，Tornado 率先支持了全新的 AIO 编程语法。本章将详细介绍 Tornado 的用法。

第 7 章 Socket. IO

Socket. IO 是一个易学易用的实时通信框架，支持浏览器与服务器之间进行实时互动。

第1章 Python AIO 库

asyncio 库最初出现在 Python 3.4 版本中，是一个使用 async/await 语法编写并发逻辑代码的库。以 asyncio 库为基础，应用层框架可以实现高性能的网络服务、数据库连接、分布式的消息队列等。

1.1 协程

4min

协程是一种比线程更轻量级的存在形式。要理解协程并不容易，我们从程序对 IO 资源的利用率开始讲起，例如开发一个服务器程序，服务器的带宽是 50Mb/s，而一个客户端的连接只占用了其中的 1Kb/s，如果服务器在处理第 1 个连接时无法处理其他连接，那么对于服务器来讲，纵使有 50Mb/s 的带宽，也没有意义，因为根本用不上，此时的 IO 资源利用率是极低的。

在传统阻塞型 IO 编程模型中，此问题的解决方案是为每个连接开辟一个独立的线程来处理该连接的所有事务，但是线程是一种重量级的存在形式，占用内存很高，在这种方案中，也许 IO 资源还没用完，内存资源就先用完了。为了避免内存资源耗尽，我们将引入线程池的概念，规定最大可开辟的线程数量，在线程池处于满载情况时如果有新的连接进来，则该连接处于等待状态，等待之前的连接完成处理，这虽然避免了内存资源耗尽的问题，但是仍然无法最有效地利用 IO 资源。

在目前科技的发展中，网速提升速度要远远大于内存提升的速度，无线通信的网络技术发展得更快，尤其是 5G 技术的发展，将我们带进了一个全新的网络世界，如果不能更有效地利用网络 IO 资源，将会是一个巨大的损失，既然硬件的发展存在瓶颈，那么是否可以从软件技术层面寻找突破呢？当然可以，那就是用一个线程去管理 IO 资源，其他所有事务的处理都基于这个线程的事件去驱动，这就解决了 IO 资源利用率瓶颈的问题。

但是在实际开发中，基于事件驱动的编程模型在处理复杂业务逻辑时会非常难以实现。举个例子，一个用户登录后列出该用户的相关信息，如图 1-1 所示，从用户发起登录请求开始，服务器程序便将前端上传的用户信息与数据库中的信息做比对，这要查询数据库，若想知道何时数据库返回结果，此时需要添加一个事件侦听器(不同编程语言中的事件侦听器的表现形式不同，在 JavaScript 中可能是一个闭包函数、在 Java 中可能是一个匿名类、在 Python 中可能是一个 lambda 表达式等)，此时需要写一层嵌套语句，在比对信息(验证密码)成功后,再次与数据库通信获取用户详细信息,此时需再添加一个事件侦听器以侦听数据库返回结果,这意味着需要再嵌套一层语句。

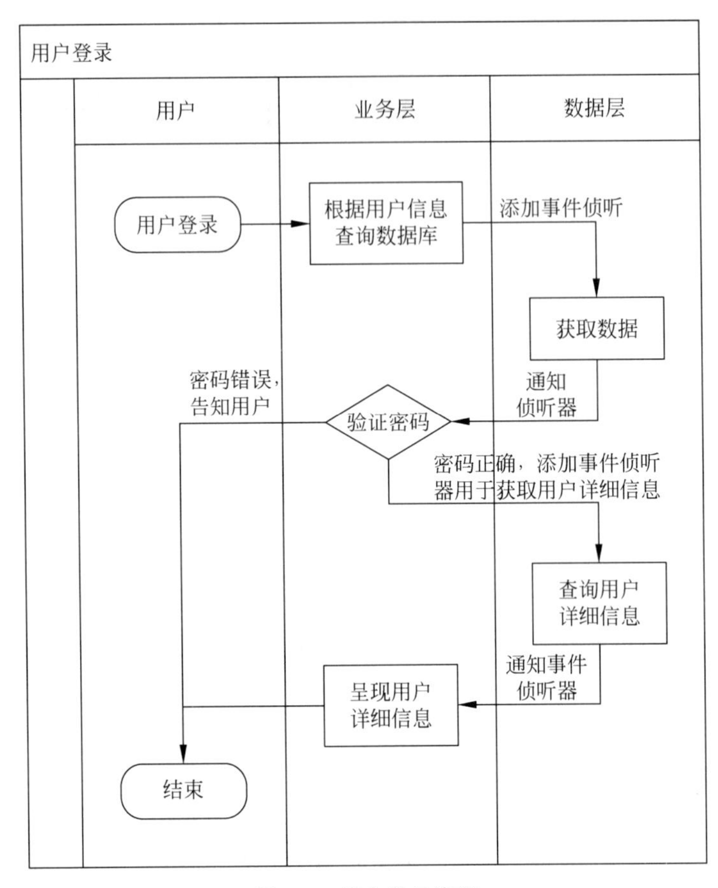

图 1-1 用户登录流程

而在实际开发工作中的业务逻辑一般会比图 1-1 中所示的业务逻辑复杂很多倍,这就意味着处理这些业务逻辑需要写很多层嵌套语句,伪代码如下:

```
do_action1(function () {
    do_action2(function () {
        do_action3();
    });
});
```

人类大脑并不擅长处理多层嵌套逻辑,而更擅长处理顺序逻辑,所以我们希望代码以顺序逻辑的形式处理问题,代码如下:

```
do_action1();
do_action2();
do_action3();
```

于是协程的概念便被提了出来,将这种事件驱动模型在编程语言层面进行支持,让处理基于事件驱动的代码逻辑写出来就像是在处理传统阻塞型 IO 的代码逻辑一样，从而大大提高开发效率，同时不会像线程那样占用大量内存资源，所以协程在提升 IO 利用率的同时又不会增加开发成本。

在 Python 语言中，协程将搭配使用 async/await 关键字进行编程。代码如下：

```
async def ajax_info(req, res):
        await res.render("user/ajax_info.html")
```

接下来我们以一个可运行的例子演示在 Python 中如何使用协程。代码如下：

```
"""
第 1 章/hello_aio/hello_aio.py
"""

import asyncio
import time


#使用 async 关键字声明一个异步函数
async def main():
    print(f"{time.strftime('%X')} Hello")
    #使用 await 关键字休眠当前协程
    await asyncio.sleep(1)
    print(f"{time.strftime('%X')} World")


asyncio.run(main())
```

在这个例子中，通过 async 关键字声明一个异步函数 main 作为协程最高层级的入口点，通过 asyncio.run 来运行。因为在协程中所有事务都是由事件驱动的，所以通过 async 关键字指明在语言层面支持这种基于事件驱动的编程模型，而这种编程模型被称为异步编程。

在异步函数中可以使用 await 关键字去等待一个可等待的对象（Awaitable），当该对象所需要完成的事务完成后，程序继续运行，所以这段程序的运行效果是首先输出一个 Hello，等待 1s 后，再输出 World，如图 1-2 所示。

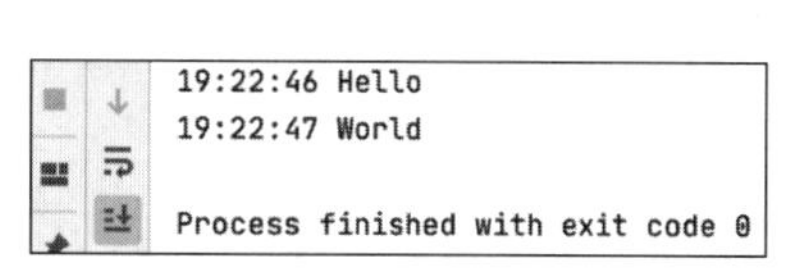

图 1-2 异步 Hello World 程序结果

1.2 任务

在协程中也存在等待的概念，就像传统阻塞型 IO 编程模型那样，那么如何支持同时处理多个事务呢？在 Python 中，通过创建任务（asyncio.create_task）的方式实现。代码如下：

```
"""
第 1 章/aio_multi_task/aio_multi_task.py
"""

import asyncio #引入 asyncio 库
import time #引入 time 库,便于格式化时间输出


#声明一个异步函数,用于运行一个独立的任务
async def task(tag, delay):
    #循环 6 次执行输出语句
    for i in range(6):
        #根据 delay 参数休眠,delay 的值以秒为单位
        await asyncio.sleep(delay)
        #按指定格式输出时间,用以标识本次输出
        print(f"[{time.strftime('%X')}]Task:{tag}, step {i}")


async def main():
    #创建第 1 个任务,不等待任务执行结束,意味着如果代码不阻塞当前的协程
    #则程序继续走下去
    asyncio.create_task(task("task1", 1))

    #创建第 2 个任务,并且使用 await 关键字等待该任务完成,如果这里不
    #等待,程序继续运行下去便运行到了程序结尾,那么整个程序在两个任务还
    #未执行完成时便退出,这并非我们所希望的结果
    await asyncio.create_task(task("task2", 2))


asyncio.run(main()) #启动 asyncio 主程序
```

第 1 个任务传入的时间间隔为 1s,而第 2 个任务传入的时间间隔为 2s,所以第 2 个任务耗时将比第 1 个任务更久,所以等待第 2 个任务能够保证我们看到两个任务都完成的完整效果。运行效果如图 1-3 所示。

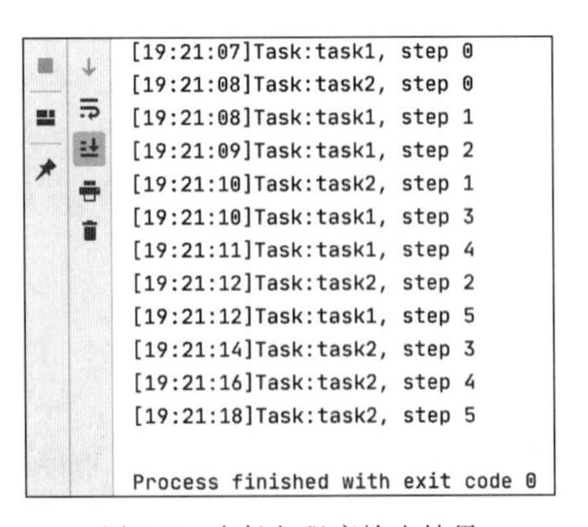

图 1-3　多任务程序输出结果

通过该效果可以看出，两个任务同时运行，并且在各自的协程中等待，在 task1 中每隔 1s 输出一句话，而在 task2 中每隔 2s 输出一句话，两个协程互不影响。

在实际开发工作中，有时我们并不知道哪一个任务最后执行完毕，所以在需要并发执行多个任务时，我们使用 asyncio.gather 函数实现，代码如下：

```python
"""
第 1 章/gather_multi_task/gather_multi_task.py
"""


import asyncio
import time


#声明一个异步函数用于运行一个独立的任务
async def task(tag, delay):
    #循环 6 次执行输出语句
    for i in range(6):
        #根据 delay 参数休眠，delay 的值以秒为单位
        await asyncio.sleep(delay)
        print(f"[{time.strftime('%X')}]Task:{tag}, step {i}")


async def main():
    #并发执行多个任务
    await asyncio.gather(
        task("task1", 1),
        task("task2", 2)
    )


asyncio.run(main())
```

既然协程有类似线程的运行效果，那么是否可以使用协程完全代替线程去处理所有的业务逻辑呢？

与异步 IO 相关的业务逻辑可以在协程中实现，但是 CPU 密集型运算类的业务逻辑则不能直接在协程中实现。例如，资源打包下载，这类逻辑往往是获取所有相关资源，并把所有资源在线压缩成一个 zip 文件，如果资源很多，压缩算法的执行时间往往会超过 10s，而这个逻辑如果运行在协程中，在压缩过程时所有协程都将处于假死状态，这意味着如果整个网站是基于 Python AIO 构建，那么整个网站将处于假死状态，所以在异步编程中，应避免直接在协程中写 CPU 密集型运算逻辑。

同理还应该避免直接在协程中写基于阻塞型 IO 的代码逻辑。

1.3 支持阻塞型 IO

通过 1.2 节的学习，我们已经了解到 CPU 密集型运算和阻塞型 IO 类的逻辑是不能直接写在协程里的，那么遇到此类问题该怎么办呢？

不用担心，既然 Python 提供了异步编程模型，那么配套的工作 Python 肯定也帮我们做好了，开发者只需调用 API。在协程中支持阻塞型 IO 代码如下：

```
"""
第 1 章/support_blocked_io/support_blocked_io.py
"""
import asyncio
import time
import concurrent.futures


#声明一个阻塞型任务
def blocked_task():
    for i in range(10):
        #为了简化代码逻辑，便于我们更加清晰地认识混合执行阻塞与非阻塞(异步)代
        #码，这里以 time.sleep 函数来模拟阻塞型 IO 逻辑的执行效果
        time.sleep(1)
        print(f"[{time.strftime('%X')}] Blocked task {i}")


#声明一个异步任务
async def async_task():
    for i in range(2):
        await asyncio.sleep(5)
        print(f"[{time.strftime('%X')}] Async task {i}")


async def main():
    #获取当前正在运行的事件循环对象，那么什么是事件循环呢?在 1.1 节中讲过协程是由
    #事件机制驱动的，而用于驱动协程的事件机制系统在 Python 中被称为
    #事件循环(Running Loop)，通过该事件循环对象可以与其他线程或进程进行沟通
    current_running_loop = asyncio.get_running_loop()

    #并发执行一个阻塞型任务和一个异步任务
    await asyncio.gather(
        #通过函数 run_in_executor 可以让指定的函数运行在特定的执行器(Executor)
        #中，例如线程池执行器(concurrent.futures.ThreadPoolExecutor)或者
        #进程池执行器(concurrent.futures.ProcessPoolExecutor)
        current_running_loop.run_in_executor(None, blocked_task),
        async_task()
    )


asyncio.run(main())
```

最终运行结果如图 1-4 所示。

从效果图可以看出，阻塞型代码逻辑与异步代码逻辑能够同时运行，而互不影响。

```
[19:18:03] Blocked task 0
[19:18:04] Blocked task 1
[19:18:05] Blocked task 2
[19:18:06] Blocked task 3
[19:18:07] Async task 0
[19:18:07] Blocked task 4
[19:18:08] Blocked task 5
[19:18:09] Blocked task 6
[19:18:10] Blocked task 7
[19:18:11] Blocked task 8
[19:18:12] Async task 1
[19:18:12] Blocked task 9

Process finished with exit code 0
```

图 1-4　支持阻塞型 IO 程序输出结果

再看 support_blocked_io.py 的 run_in_executor 函数的第 1 个参数，我们传进 None，那么代码在运行时内部究竟执行什么操作呢？我们可以通过分析 Python SDK 中的源码来了解，代码如下：

```
"""
第 1 章/support_blocked_io/base_events_sample_code.py
"""


def run_in_executor(self, executor, func, *args):
    self._check_closed()
    if self._debug:
        self._check_callback(func, 'run_in_executor')
    if executor is None:
        executor = self._default_executor
        if executor is None:
            executor = concurrent.futures.ThreadPoolExecutor()
            self._default_executor = executor
    return futures.wrap_future(
        executor.submit(func, *args), loop=self)
```

通过代码 base_events_sample_code.py 可以看出，如果我们给 executor 参数传入 None，那么系统将默认创建一个线程池执行器来执行代码。如果开发者需要自定义线程池的相关参数，则可以传入自定义的线程池执行器，所以可以对代码 support_blocked_io.py 进行修改，代码如下：

```
"""
第 1 章/support_blocked_io/support_blocked_io_modified.py
"""
import asyncio
import time
import concurrent.futures


#声明一个阻塞型任务
```

```
def blocked_task():
    for i in range(10):
        #为了简化代码逻辑,便于我们更加清晰地认识混合执行阻塞与非阻塞(异步)代
        #码,这里以 time.sleep 函数来模拟阻塞型 IO 逻辑的执行效果
        time.sleep(1)
        print(f"[{time.strftime('%X')}] Blocked task {i}")


#声明一个异步任务
async def async_task():
    for i in range(2):
        await asyncio.sleep(5)
        print(f"[{time.strftime('%X')}] Async task {i}")


async def main():
    #创建一个线程池执行器,该执行器所允许的最大线程数是 5
    executor = concurrent.futures.ThreadPoolExecutor(max_workers = 5)

    #获取当前正在运行的事件循环对象,那么什么是事件循环呢?在 1.1 节中讲过协程是由
    #事件机制驱动的,而用于驱动协程的事件机制系统在 Python 中被称为
    #事件循环(Running Loop),通过该事件循环对象可以与其他线程或进程进行沟通
    current_running_loop = asyncio.get_running_loop()

    #并发执行一个阻塞型任务和一个异步任务
    await asyncio.gather(
        #通过函数 run_in_executor 可以让指定的函数运行在特定的执行器(Executor)
        #中,例如线程池执行器(concurrent.futures.ThreadPoolExecutor)或者
        #进程池执行器(concurrent.futures.ProcessPoolExecutor)
        current_running_loop.run_in_executor(executor, blocked_task),
        async_task()
    )


asyncio.run(main())
```

1.4 支持 CPU 密集型运算

3min

支持 CPU 密集型运算与支持阻塞型 IO 的操作方式大同小异,不同之处就是把线程池执行器换成进程池执行器,这里我们以计算圆周率为例演示如何支持 CPU 密集型运算,代码如下:

```
"""
第 1 章/cpu_bound_computing/cpu_bound_computing.py
"""

import asyncio
```

```
import concurrent.futures

from numba import jit
import time


#一种计算圆周率算法
@jit #通过 JIT 加速,更高效地执行 CPU 密集型运算
def compute_pi():
    #因为 Python 程序执行速度是比较慢的(相对于所有编译型编程语言和大多数解
    #释型编程语言),所以在进行 CPU 密集型运算时显得更加吃力,因此我们在这里进行了
    #JIT 加速,其原理就是在执行这段代码时先将代码编译成机器码再执行,这样可以大大提高
    #程序的运行速度.如果该程序在计算机上运行所需时间仍然较长,可适当调低
    #count 的数值
    count = 100000
    part = 1.0 / count
    inside = 0.0
    for i in range(1, count):
        for j in range(1, count):
            x = part * i
            y = part * j
            if x * x + y * y <= 1:
                inside += 1
    pi = inside / (count * count) * 4
    return pi


async def print_pi(pool):
    print(f"[{time.strftime('%X')}] Started to compute PI")
    #将计算圆周率(CPU 密集型)的代码交给进程池执行器执行
    pi = await asyncio.get_running_loop().run_in_executor(
        pool,
        compute_pi
    )
    print(f"[{time.strftime('%X')}] {pi}")


async def task():
    for i in range(5):
        print(f"[{time.strftime('%X')}] Step {i}")
        await asyncio.sleep(1)


async def main():
    #声明一个进程池执行器对象,与线程池执行器一样只需声明一次便可以在多处使用
    pool = concurrent.futures.ProcessPoolExecutor()

    await asyncio.gather(
        #将线程池对象 pool 传入 print_pi 函数,由 print_pi 函数执行 CPU 密集
        #型代码逻辑,并且我们将 CPU 密集型代码与异步代码并行执行
```

```
        print_pi(pool),
        task()
    )


if __name__ == '__main__':
    """
这里需要判断只有在文件名为 __main__ 时才会执行主程序,为了避免在创建子
进程时重复运行主程序而产生错误
    """
    asyncio.run(main())
```

本示例依赖第三方库 numba,需要使用 pip 单独安装。代码运行结果如图 1-5 所示。

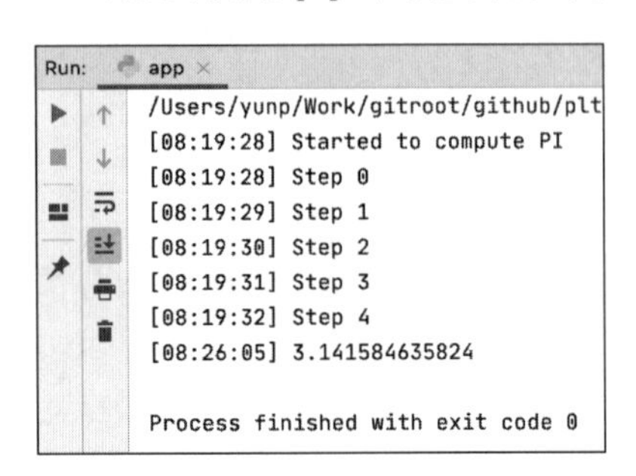

图 1-5 支持 CPU 密集型运算程序输出结果

从结果可以看出,CPU 密集型运算与协程代码同时运行,而互不影响。

扩展阅读:圆周率算法

圆周率算法有很多种,为了便于理解,在此介绍一种非常直观的算法。

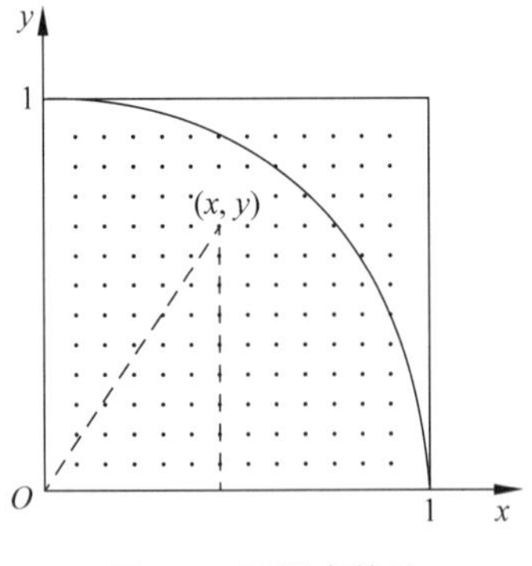

图 1-6 圆周率算法

如图 1-6 所示,将 1 进行无限均等分割,由此产生的所有坐标点将均匀分布在边长为 1 的正方形内,所有点的数量可认为是该正方形的面积(S_s),则满足 $x^2+y^2\leqslant 1$ 的所有点数量可认为是该扇形的面积(S_f)。

已知扇形半径(r)等于正方形边长,则 $S_s=r^2$,$S_f=\pi r^2/4$,可推出 $\pi=S_f * 4/S_s$,代码如下:

```
def compute_pi():
    count = 100000
    part = 1.0 / count
    inside = 0.0
    for i in range(1, count):
        for j in range(1, count):
            x = part * i
            y = part * j
            if x * x + y * y <= 1:
                inside += 1
    pi = inside / (count * count) * 4
    return pi
```

从代码中可以看出,count 值越大,则分割得越细致,所得出的圆周率就越精确,同等运算能力下所需要的运算时间则更久。

1.5 文件异步 IO

3min

在 Python 3.8 中尚未提供文件异步 IO API,但在 1.3 节我们已经学习过如何支持阻塞型 IO,接下来演示如何在协程中生成文件,代码如下:

```
"""
第 1 章/aiofile/write_file.py
"""
import asyncio #引入 asyncio 模块


async def main():
    #获取事件循环
    loop = asyncio.get_running_loop()
    #将阻塞型 IO 的 open 函数运行在线程池执行器(ThreadPoolExecutor)中,
    #以写入字符串的方式打开文件 data.txt
    f = await loop.run_in_executor(None, open, "data.txt", 'w')
    #将数据写入文件中
    await loop.run_in_executor(None, f.write, "aio file")
    #关闭文件
    await loop.run_in_executor(None, f.close)


if __name__ == '__main__':
    asyncio.run(main())
```

在每一次要执行阻塞型 IO 的操作时,都需要使用 run_in_executor 函数进行包装。在实际开发工作中,这样写代码是很烦琐的,我们应该基于这个原理封装一个用于操作文件异步 IO 的库,将此库命名为 aiofile.py,代码如下:

```
"""
第 1 章/aiofile/aiofile.py
"""

import asyncio
from typing import Any


class AsyncFunWrapper:

    def __init__(self, blocked_fun) -> None:
        super().__init__()

        #记录阻塞型 IO 函数,便于后续调用
        self._blocked_fun = blocked_fun

    def __call__(self, *args):
        """
重载函数调用运算符,将阻塞型 IO 的调用过程异步化,并返回一个可
等待对象(Awaitable)。通过重载运算符实现包装逻辑的好处是不用
一个一个去实现阻塞型 IO 的所有成员函数,从而大大节省了代码量
        """
        return asyncio.get_running_loop().run_in_executor(
            None,
            self._blocked_fun,
            *args
        )


class AIOWrapper:
    def __init__(self, blocked_file_io) -> None:
        super().__init__()
        #在包装器对象中记录阻塞型 IO 对象,外界通过包装器调用其成员
        #函数时,事实上分成两步进行
        #第一步,获取指定的成员,而该成员是一个可被调用的
        #对象(Callable)
        #第二步,对该成员进行调用
        self._blocked_file_io = blocked_file_io

    #重载访问成员的运算符
    def __getattribute__(self, name: str) -> Any:
        """
在外界通过包装器(AIOWrapper)访问成员操作时,创建一个异步
函数包装器(AsyncFunWrapper),其目的是将函数调用过程异步化
        """
        return AsyncFunWrapper(
            super().__getattribute__(
                "_blocked_file_io"
            ).__getattribute__(name)
        )
```

```
async def open_async( * args) -> AIOWrapper:
    """
当外界调用该函数时,将返回一个包装器(AIOWrapper)对象,该包装器
包装了一个阻塞型 IO 对象
    """
    return AIOWrapper(
        #通过 run_in_executor 函数执行阻塞型 IO 的 open 函数,并转发外
        #界传入的参数
        await asyncio.get_running_loop().run_in_executor(
            None, open, * args
        )
    )
```

基于 aiofile.py 的实现,我们可以将 write_file.py 的代码简化,代码如下:

```
"""
第 1 章/aiofile/write_with_aiofile.py
"""
import asyncio

import aiofile #引入已实现的 aiofile 模块


async def main():
    #通过异步方式以写文本模式打开 data.txt 文件
    f = await aiofile.open_async("data.txt", "w")
    #向文件中写入数据
    await f.write("aio file")
    #关闭文件
    await f.close()


if __name__ == '__main__':
    asyncio.run(main())
```

运行该代码将生成 data.txt 文件,如图 1-7 所示。

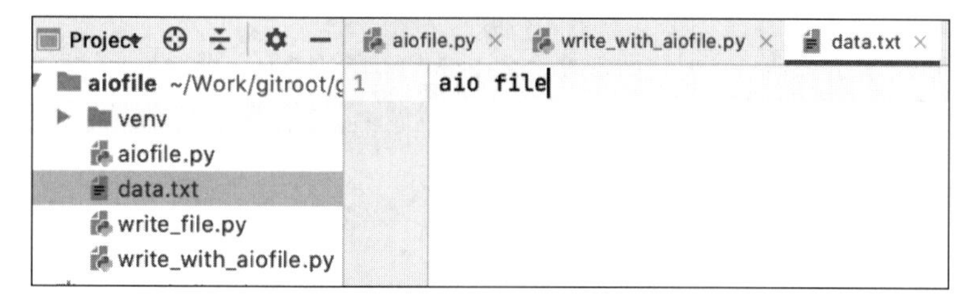

图 1-7 生成的文件内容

1.6 异步 Socket 服务器

4min

在处理实际开发中的业务逻辑时，网络 IO 往往比文件 IO 要复杂得多，虽然 Python 暂时没有推出基于异步 IO 的文件系统 API，但将来一定会推出配套的基于异步 IO 的网络 API，从而简化网络 IO 的使用难度。

接下来创建一个简单的 Socket 服务器，用于向连接终端以间隔 1s 发送 5 段数据，之后关闭连接，代码如下：

```
"""
第 1 章/aio_socket_server/aio_socket_server.py
"""

import asyncio


async def handle_echo(reader, writer):
    """
    handle_echo 向客户端发送 5 段数据，时间间隔为 1s，之后关闭客户端连接
    :param reader: reader 是一个 StreamReader 对象，用于读取客户端传来的数据
    :param writer: writer 是一个 StreamWriter 对象，用于向客户端发送数据
    """

    #向客户端输出 5 次数据
    for i in range(1, 6):
        writer.write(f'Count {i}\n'.encode('utf-8'))
        #发送并等待数据发送成功
        await writer.drain()
        #休眠 1s
        await asyncio.sleep(1)

    writer.close() #关闭客户端连接


async def main():
    #声明端口
    port = 8888
    #启动服务器并注册一个回调函数用于侦听客户端连接，当有一个新的
    #客户端建立连接时，将触发 handle_echo 函数
    server = await asyncio.start_server(
        #客户端连接的回调函数
        handle_echo,
        #指定端口
        port=port
    )
    print(f'Serving on port {port}')
```

```
    async with server:
        #开始接受连接
        await server.serve_forever()


if __name__ == '__main__':
    try:
        asyncio.run(main())
    except KeyboardInterrupt: #用户强制退出时捕获该异常
        print("User stopped server")
```

现在启动该服务器,输出结果如图 1-8 所示。

我们可以使用 telnet 客户端连接该服务器以便进行测试,结果如图 1-9 所示。

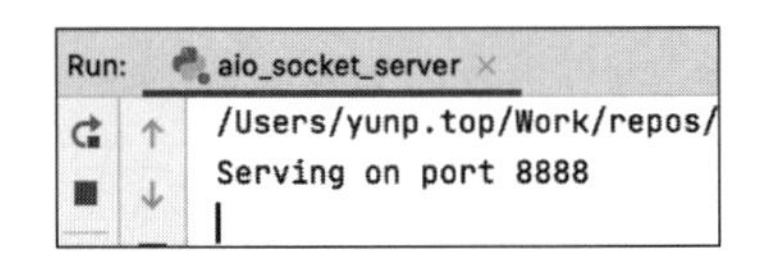

图 1-8 服务器运行结果

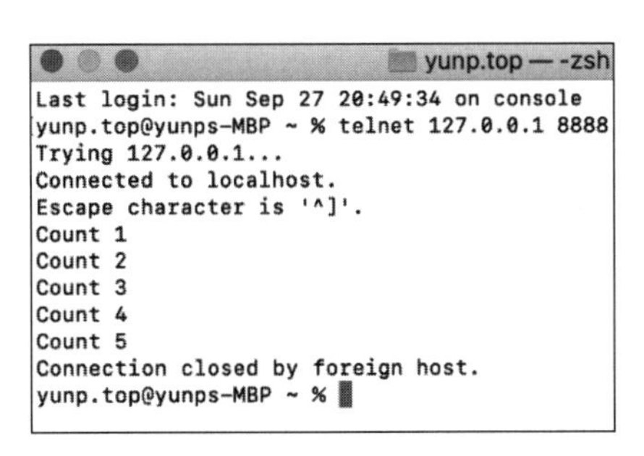

图 1-9 telnet 客户端连接服务器

1.7 异步 Socket 客户端

5min

本节演示如何使用 Socket 客户端,代码如下:

```
"""
第 1 章/aio_socket_client/aio_socket_client.py
"""

import asyncio
import time


async def main():
    #通过 asyncio.open_connection 函数创建一个到
    #本机 8888 端口的连接,可用于连接 aio_socket_server.py
    #所启动的服务器
    reader, writer = await asyncio.open_connection(
        '127.0.0.1', 8888
    )

    #通过循环不断地一行一行读取数据,并将读到的数据输出到终端
    #在读到数据结尾时跳出循环
```

```
        while not reader.at_eof():
            #读取一行数据
            data = await reader.readline()
            print(f"[{time.strftime('%X')}] Received: {data}")


asyncio.run(main())
```

在运行该代码之前需先启动 aio_socket_server.py 服务器，本例运行效果如图 1-10 所示。

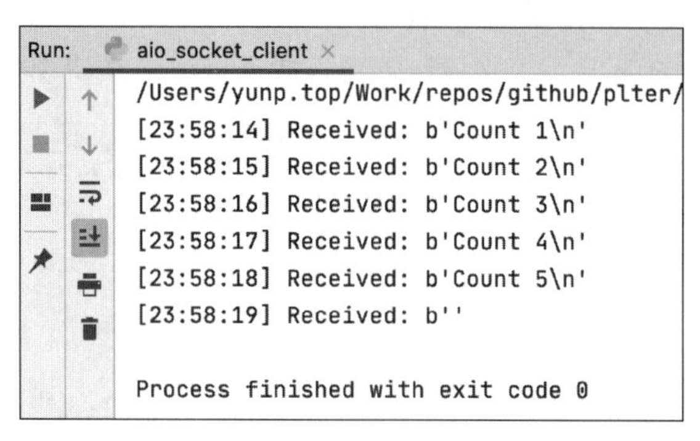

图 1-10　Python AIO 客户端连接服务器

1.8　异步 HTTP 客户端

5min

与 Socket 客户端相比，HTTP 客户端更加常用。例如，在服务器框架设计中，通常会将数据层单独架设，以 REST API 的方式提供给内部应用服务器调用，那么此时就需要应用服务器通过 HTTP 的方式去请求数据服务器。

因为 HTTP 是基于 Socket 的，所以可以以同样的方式建立连接，按照 HTTP 的标准去发送和读取数据。

当使用浏览器打开一个网页时，所进行的操作实际上是由浏览器与指定域名的服务器之间建立了一个 Socket 连接，并向服务器发送了一段包括请求信息的数据，服务器端解析该数据，并根据浏览器的请求返回指定的数据。

一个浏览器在请求中向服务器传输的原始数据如图 1-11 所示，该截图来自 Firefox 浏览器的开发者工具控制台。

查看请求头的原始数据信息，可以看到其中第 1 行内容是 GET /app/index HTTP/1.1，指的是向服务器端发送一个 GET 方式的请求，该请求要求呈现页面 /app/index，所使用的 HTTP 的版本是 1.1。从第 2 行开始是 HTTP 协议头信息，可包括很多信息，如浏览器可接受的语言、可接受的压缩方式、Cookie 信息等，最终请求头以\r\n\r\n 结束，服务器端一旦读到 \r\n\r\n，便认定该请求已结束，可以开始根据该请求向前端发送数据(关于 HTTP 的标准，可参考 Mozilla 所提供的文档 developer.mozilla.org)。

图 1-11 浏览器原始请求数据截图

接下来以一个示例演示如何加载 yunp.top 首页的内容,代码如下:

```
"""
第 1 章/aio_http_client/aio_http_client.py
"""
import asyncio


async def main():
    # open_connection 函数有两个返回值 reader 和 writer,其中 reader 用
    # 于读取服务器的数据,writer 用于向服务器发送数据
    reader, writer = await asyncio.open_connection("yunp.top", "80")
    # HTTP 协议的第一行,指定请求资源及 HTTP 版本
    writer.write(b'GET / HTTP/1.0\r\n')
    # 发送 HTTP 协议头结尾标识
```

```
    writer.write(b'\r\n')

    #write 函数会尝试立即向 Socket 连接发送数据,但是也有可能失败,最常见的原因
    #是当前 IO 资源被占用,失败时数据以队列形式暂存于缓冲区直到可以被再次发送,为
    #确保数据完全发送成功之后再进行后续操作,通常使用 drain 函数来等待数据发送
    #完毕,如果在调用该函数前数据已经发送完毕,则该函数立即返回结果
    await writer.drain()
    result = await reader.read()
    print(result.decode("utf-8"))


if __name__ == '__main__':
    asyncio.run(main())
```

该示例中,我们实现了一个最简单的请求,不包括任何附加的头信息,只有第1行内容指定以 GET 方式请求网站根路径中资源,所使用的 HTTP 版本是 1.0,因为 HTTP 1.0 不包括长连接,所以该请求不需要指定连接(Connection)的方式,服务器端根据请求返回数据后立即关闭该连接,这就是一个完整的 HTTP 请求过程。

如果在使用 HTTP 1.1 的时候需要建立短连接,则可以手动指定 Connection 的值为 close,代码如下:

```
"""
第1章/aio_http_client/aio_http_client_sc.py
"""
import asyncio


async def main():
    #open_connection 函数有两个返回值 reader 和 writer,其中 reader 用
    #于读取服务器的数据,writer 用于向服务器发送数据
    reader, writer = await asyncio.open_connection("yunp.top", "80")
    #HTTP 的第一行,指定请求资源及 HTTP 版本
    writer.write(b'GET / HTTP/1.1\r\n')
    #HTTP 协议头,Host 指定请求的主机,注意在冒号后面必须有一个空格
    writer.write(b"Host: yunp.top\r\n")
    #HTTP 协议头,Connection 指定连接服务器的方式
    writer.write(b"Connection: close\r\n")
    #发送协议头结尾标识
    writer.write(b'\r\n')

    #write 函数会尝试立即向 Socket 连接发送数据,但是也有可能失败,最常见的原因
    #是当前 IO 资源被占用,失败时数据以队列形式暂存于缓冲区直到可以被再次发送,为
    #确保数据完全发送成功之后再进行后续操作,通常使用 drain 函数来等待数据发送
    #完毕,如果在调用该函数前数据已经发送完毕,则该函数立即返回结果
    await writer.drain()
    result = await reader.read()
    print(result.decode("utf-8"))
```

```
if __name__ == '__main__':
    asyncio.run(main())
```

运行结果如图 1-12 所示。

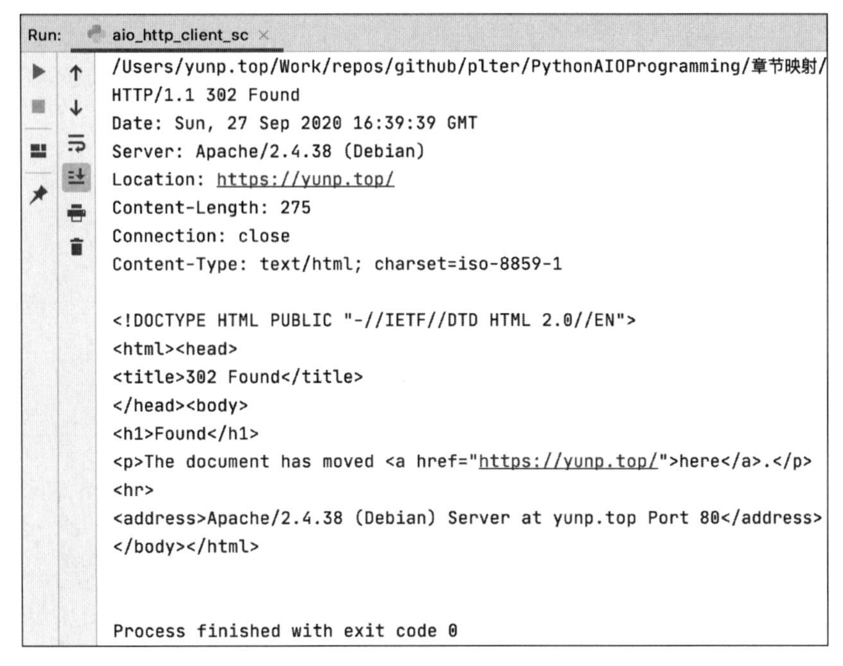

图 1-12 HTTP 客户端请求结果

如果在浏览器中访问同样的地址，则会自动解析服务器的返回结果，并进行页面渲染。

在 HTTP 请求中，请求方式除了 GET，还有 POST、HEAD、PUT 等，其功能还有上传数据、上传文件、根据头信息判断是否使用本地缓存等。可以看出，如果要基于 Socket 实现一个完整的 HTTP 客户端，工作量还是很大的，在实际开发工作中，往往直接使用第三方已经实现好的 HTTP 客户端库。当然如果你有兴趣也可以自己实现一遍，我相信对于提升你的编程能力肯定有巨大的帮助。

1.9 异步 HTTP 服务器

3min

HTTP 是基于 Socket 的，所以只需对 Socket 服务器进行改进便可实现 HTTP 服务器。由于 HTTP 的内容非常庞大，如果把所有的标准都实现，工作量非常巨大，而目前已经有许多团队完成了该项任务，我们没有必要重复这些已经完成的任务，但是掌握如何实现 HTTP 服务器这项技术还是非常有必要的。

在这里以一个简单的实例来演示最核心的功能，代码如下：

```
"""
第 1 章/aio_http_server/aio_http_server.py
"""
```

```
import asyncio


async def handle_connection(reader, writer):
    content = b'<html>'\
              b'<head>'\
              b'<title>Title</title>'\
              b'</head>'\
              b'<body>'\
              b'Hello World'\
              b'</body>'\
              b'</html>'
    #设置 http 响应的状态,200 表示成功
    writer.write(b"HTTP/1.0 200 OK\r\n")
    #指定 http 响应内容的长度
    writer.write(f"Content-Length: {len(content)}\r\n".encode('utf-8'))
    #指定 http 响应内容的格式
    writer.write(b"Content-Type: text/html\r\n")
    #指定头部信息结束,后面的信息就是内容。HTTP 标准规定,当读到\r\n\r\n 时,便认定是
#头部结束
    writer.write(b"\r\n")
    #发送响应内容
    writer.write(content)
    #等待发送完成
    await writer.drain()
    #关闭连接
    writer.close()


async def main():
    async with (
            #配置服务器端口
            await asyncio.start_server(handle_connection, port=8888)
    ) as server:
        #启动服务器
        await server.serve_forever()


asyncio.run(main())
```

创建 HTTP 服务器的方式与创建 Socket 服务器并没有太大区别,区别在于处理连接上,handle_connection 函数实现了一个简单地向浏览器发送 HTTP 响应消息的逻辑。在启动服务器后使用浏览器访问的效果如图 1-13 所示。

127.0.0.1:8888

Hello World

图 1-13 浏览器访问 HTTP 服务器效果图

5min

1.10 子进程

Python 语言有一个重要的用途，就是可以将其当成脚本使用，在推出异步 IO 的同时绝对不能失去这个重要的用途，所以 Python 配套推出了完整的、非常易用的、基于异步 IO 的子进程交互 API。

接下来以一个列出当前目录下所有文件的示例来演示调用本机命令，代码如下：

```
"""
第 1 章/sub_process/sub_process_example.py
"""

import asyncio


async def main():
    # 创建进程,用于执行系统命令 ls
    p = await asyncio.create_subprocess_shell("ls")
    # 等待该进程执行完毕
    await p.wait()

asyncio.run(main())
```

在 Windows 系统中没有 ls 命令，需要使用 dir 替换，则 Windows 系统中的代码如下：

```
"""
第 1 章/sub_process/sub_process_example_win.py
"""

import asyncio


async def main():
    # 创建进程,用于执行系统命令 ls
    p = await asyncio.create_subprocess_shell("dir")
    # 等待该进程执行完毕
    await p.wait()


asyncio.run(main())
```

为什么要等待该进程执行结束呢？因为代码是异步的，在不等待的情况下，子进程可能还没有启动，主进程便已经结束了，那么子进程也会随主进程一起结束，这种情况下将无法看到运行子进程的任何效果。

我们先来看一下在苹果系统上的运行效果，如图 1-14 所示。

再来看一下在 Windows 系统上的运行效果，如图 1-15 所示。

在苹果系统上没有问题，但是在 Windows 系统上出现了乱码，这是因为在苹果系统上所有命令的输出信息统一使用 UTF-8 编码方式，而在简体中文 Windows 系统上的命令默认使用 GB 2312 的编码方式进行输出，而 PyCharm 环境中使用 UTF-8 的方式来呈现，所以用户看到的是乱码。

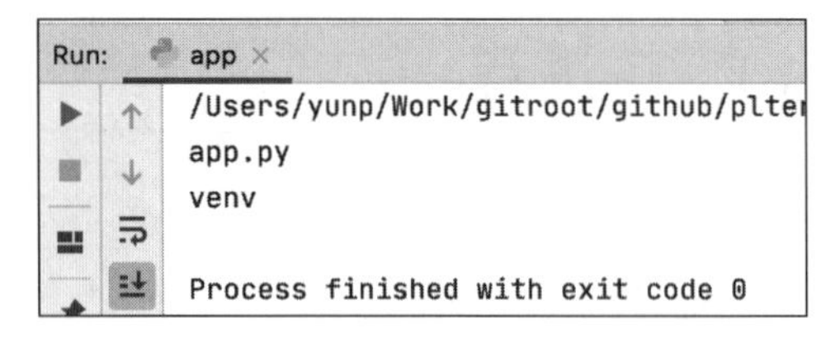

图 1-14 在苹果系统上运行 ls 子进程结果

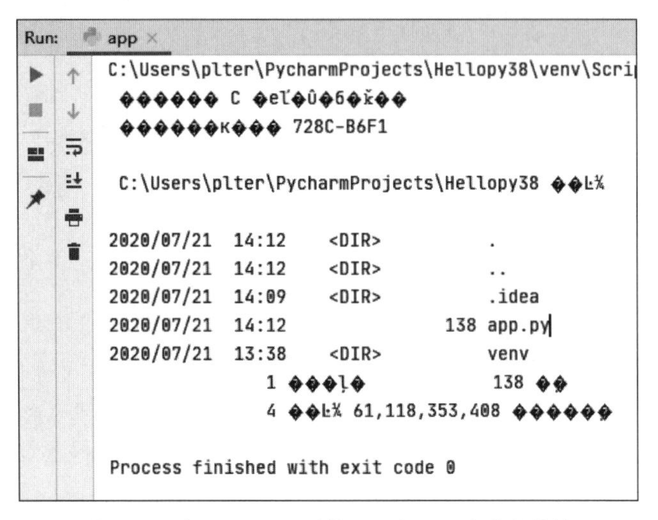

图 1-15 在 Windows 系统上运行 dir 子进程结果

这个问题可以不用去解决，因为 Python 脚本程序一般会直接运行于终端里，而在 Windows 系统的终端里直接运行该程序是没有问题的。图 1-16 所示的是在 Windows 系统的终端里运行该代码的结果。

```
命令提示符

(venv) C:\Users\plter\PycharmProjects\Hellopy38>python app.py
 驱动器 C 中的卷没有标签。
 卷的序列号是 728C-B6F1

 C:\Users\plter\PycharmProjects\Hellopy38 的目录

2020/07/21  14:23    <DIR>          .
2020/07/21  14:23    <DIR>          ..
2020/07/21  14:23    <DIR>          .idea
2020/07/21  14:23               138 app.py
2020/07/21  13:38    <DIR>          venv
               1 个文件            138 字节
               4 个目录 61,116,973,056 可用字节
```

图 1-16 在 Windows 系统终端里运行结果

当然有些程序员有强迫症，总觉得是问题就得解决。这也有道理，其实可以将此看作对于知识的渴求，而这种渴求也会引导着我们学会更多的知识。

因为已知 Windows 系统的默认编码方式是 GB 2312，我们可以将子进程的输出信息重

定向到 asyncio 中，之后通过 Python 程序以指定的编码方式来解析成字符串，这样便可以正确呈现出结果，代码如下：

```
"""
第 1 章/sub_process/decode_sub_process_out.py
"""

import asyncio

async def main():
    #创建进程,用于执行系统命令 ls
    p = await asyncio.create_subprocess_shell(
        "dir",
        stdout = asyncio.subprocess.PIPE
    )

    #使用 communicate 函数与子进程通信,并等待子进程结束,在子进程
    #结束后获得子进程的输出信息
    stdout, stderr = await p.communicate()
    #将信息以 gbk(GB 2312 是简体中文,gbk 同时兼容简体中文与繁体中文)
    #的编码方式解码成字符串并输出
    print(stdout.decode("gbk"))


asyncio.run(main())
```

运行的效果如图 1-17 所示。

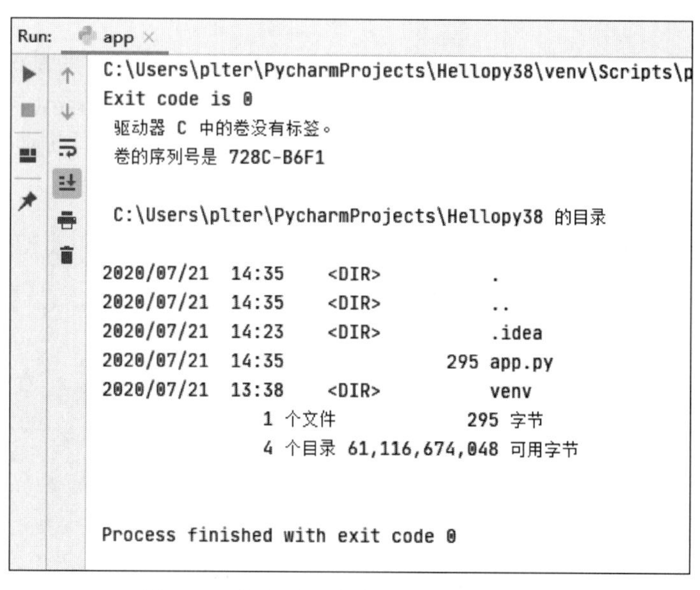

图 1-17　指定子程序输出数据的编码方式

有些命令的输出信息并非一次性的，例如 ffmpeg 命令的转码过程是不断地输出新的数据，这就要求主进程可以侦听数据输出并实时处理。若适应这种需求，我们则可以将代码改

进，代码如下：

```
"""
第 1 章/sub_process/read_sub_process_until_end.py
"""

import asyncio


async def main():
    #创建进程，用于执行系统命令 ls
    p = await asyncio.create_subprocess_shell(
        "ls",
        #将输出信息重定向到子进程管道，而不是直接输出到系统终端
        #这样便于后续通过该子进程直接读取数据
        stdout = asyncio.subprocess.PIPE
    )

    #在没有结束之前一直读取数据，eof 全称为 End of file，意为文件结尾
    #指该数据的结尾
    while not p.stdout.at_eof():
        #通过子进程的标准输出管道读取一行内容
        line = await p.stdout.readline()
        if line:
            print(line)


asyncio.run(main())
```

有时需要执行一个可执行程序，但是该程序并不在当前系统环境变量里，那么该如何执行呢？这时只需知道该程序的路径，便可执行它，使用的是 asyncio.create_subprocess_exec 函数。接下来以执行本地的 ffmpeg 为例演示如何使用 Python 异步 IO 直接执行可执行应用程序，代码如下：

```
"""
第 1 章/sub_process/sub_process_exec.py
"""

import asyncio


async def main():
    #创建进程，用于执行系统命令 ls
    p = await asyncio.create_subprocess_exec(
        #文件路径可以换成任何本机真实存在的可执行程序的全路径
        "/Users/yunp.top/Work/tools/ffmpeg",
        #将输出信息重定向到子进程管道，而不是直接输出到系统终端
        #这样便于后续通过该子进程直接读取数据
```

```
        stdout = asyncio.subprocess.PIPE,
        stderr = asyncio.subprocess.PIPE
    )

    # 在没有结束之前一直读取数据,eof 全称为 End of file,意为文件结尾
    # 指该数据的结尾
    while not p.stderr.at_eof():
        # 通过子进程的标准输出管道读取一行内容
        line = await p.stderr.readline()
        if line:
            print(line)


asyncio.run(main())
```

在该示例中,有一点需要注意,ffmpeg 的输出信息是通过 stderr(标准错误输出)输出的,所以在读取数据时通过 stderr 进行。

第2章 Docker 工具

Docker 是一个开源的容器化工具，其基础功能与服务是免费的。

长期以来开发者在部署服务端应用程序时都是极为痛苦的，这是因为服务器端的应用程序结构往往极为复杂，配置其运行环境需要花费大量的精力，稍有不慎就会出现问题。如果一家运营中的网站出现了问题，那对于网站来讲，损失可能是不可估量的。

此外，开发者在迁移服务器端应用程序时也是非常痛苦的，有些超大的网站光配置服务器端的运行环境就可能需要花几天的时间，而每迁移一次都要重新配置一次，这还不算那些集群服务器的配置。想想还真是应了那四个字——人肉运维。

Docker 的出现是针对性地解决这些痛点的，服务器应用编排好，不管是部署在哪里，还是迁移到哪里，均可一键启动，这样便大大节省了部署的时间，提高了工作效率，同时减少了对运维这个工种人员的需求，因为开发人员必须编排 Docker 服务，而在 Docker 服务编排完成后也就不需要运维了，对于创新型的互联网企业来说，建设团队的成本降低了。

8min

2.1 安装 Docker 及 Docker compose

Docker 的安装非常人性化，在 Windows 和 Mac OS X 平台都有官方提供的安装包，可以一键安装。打开网址 https://www.docker.com/get-started，下载安装包后安装即可，如图 2-1 所示。

在 Windows 平台安装时有一点需要注意，需要启用 WSL 2 功能，如图 2-2 所示。

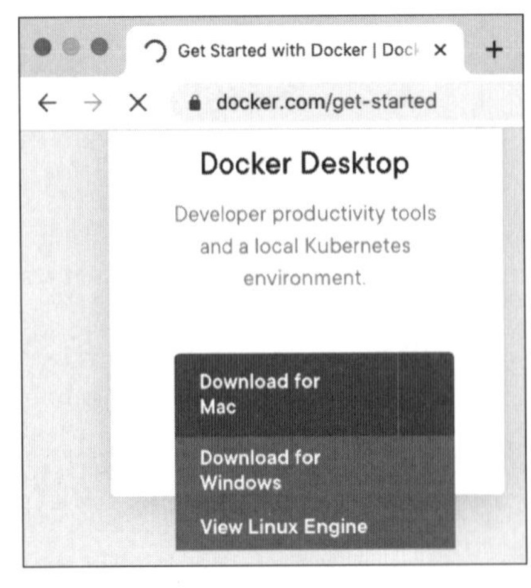

图 2-1 下载 Docker

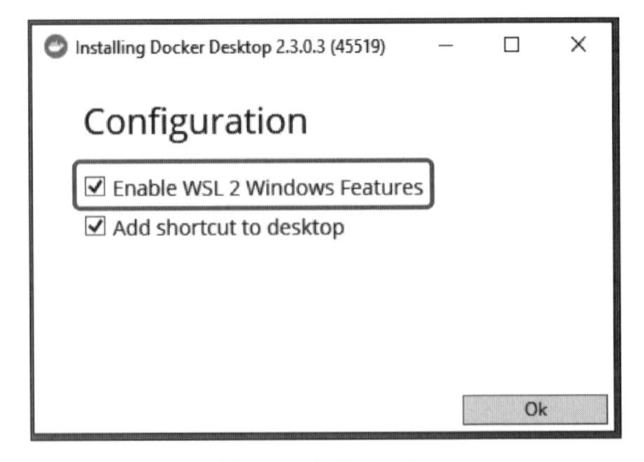

图 2-2 安装 Docker

在 Linux 平台安装稍微复杂一些，因为 Linux 有很多种发行版，所以 Docker 官方为常用的几个发行版分别提供了安装文档，开发者按照文档便可顺利安装。打开网址 https://docs.docker.com/engine/install/，选择与自己的 Linux 发行版匹配的文档进行参考，如图 2-3 所示。

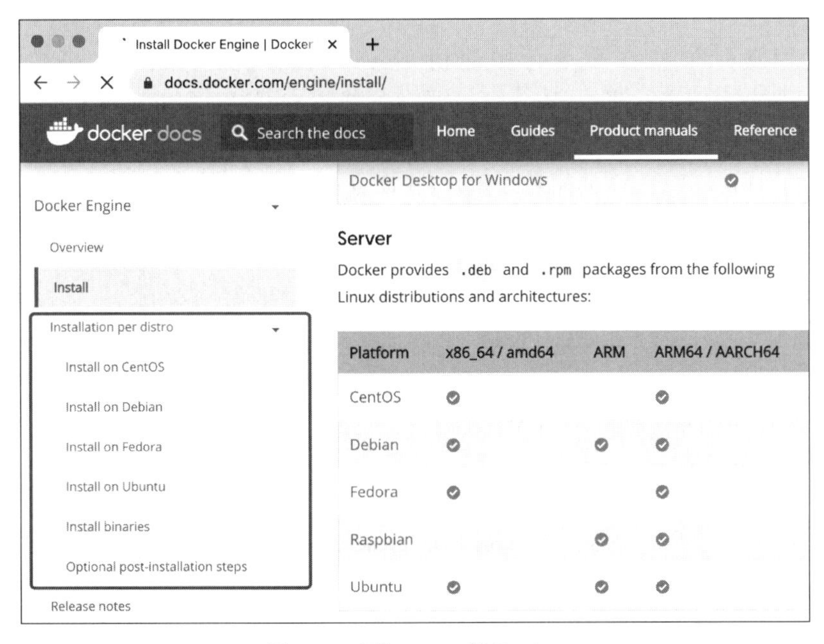

图 2-3　安装 Linux 版 Docker

如果你使用 Mac 或者 Windows 的安装包直接安装 Docker，那么在安装时会自动安装配套工具 docker-compose，但是如果你在 Linux 平台进行安装，则需要自己手动安装 docker-compose。参考文档 https://docs.docker.com/compose/install/，如图 2-4 所示。

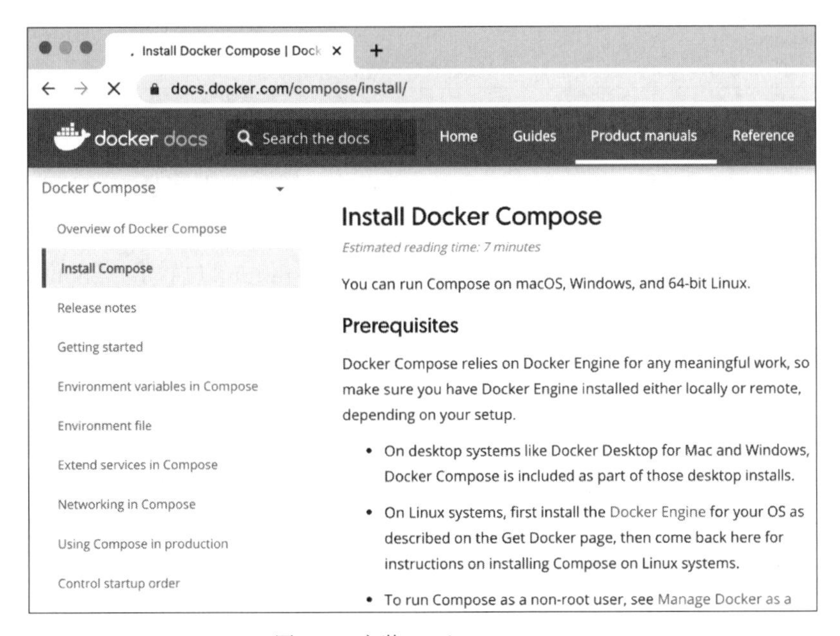

图 2-4　安装 Docker Compose

Docker 官方的文档非常详细，在安装完成后，可以通过阅读 Docker 官方文档来学习 Docker 的基本操作。

8min

2.2 使用 Docker 命令

首先我们演示如何在命令行中使用 Docker 镜像，为了照顾大多数读者，我们以 Windows 平台为例进行演示，Windows 平台的终端有 CMD 和 PowerShell，我们统一使用 PowerShell 来执行。如果你使用的是苹果计算机，则可以在终端中输入同样的命令；如果你使用的是 Linux 平台，需使用超级用户（在命令前加 sudo）来运行这些命令。

打开网址 https://hub.docker.com/，搜索 Python，会出现如图 2-5 所示界面。

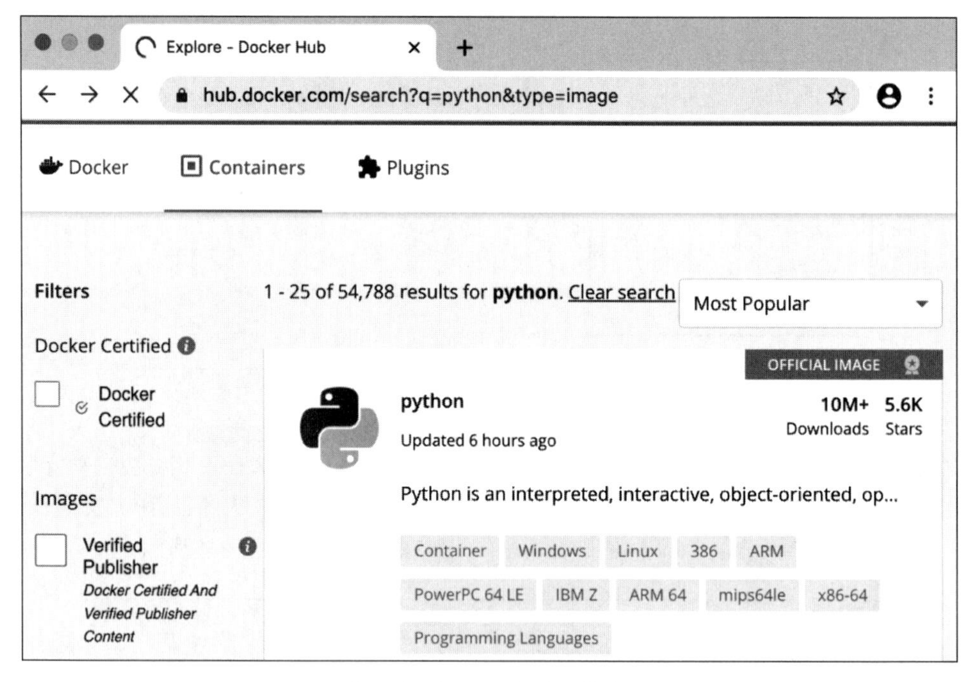

图 2-5　Python Docker 镜像

打开 Python 链接，将看到所有版本，我们可以选择一个占用空间最小的版本 3.8-alpine，如图 2-6 所示。

在 PowerShell 执行命令 docker run-ti python:3.8-alpine 以启动 Python 环境，其中-ti 参数指的是该环境拥有可交互的终端，如图 2-7 所示。

当使用 exit 函数退出该 Python 环境后，该容器还是存在的，只是处于停止状态，仍可通过 docker ps-a 命令查看，如图 2-8 所示。

这里需要理解什么是镜像、什么是容器，镜像可以理解为操作系统安装包，而容器就是使用镜像安装之后的可运行的操作系统。一个镜像可以被用来同时启动多个容器，每个容器在运行过程中也可以进行不同的配置，如果要将容器的新配置保留下来，可使用 docker commit 命令将容器保存成镜像，但一般情况下会采用编写 Dockerfile 脚本的方式创建镜像，这种方式便于一键生成。

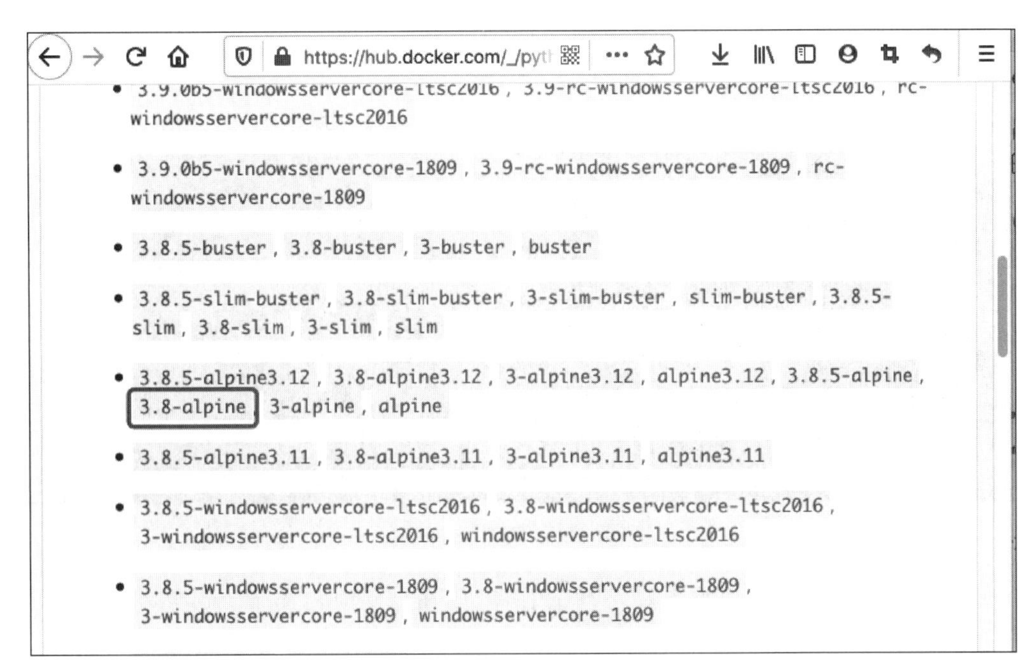

图 2-6　选择 Python:3.8-alpine

```
Windows PowerShell
Windows PowerShell
Copyright (C) Microsoft Corporation. All rights reserved.

Try the new cross-platform PowerShell https://aka.ms/pscore6

PS C:\Users\yunp> docker run -ti python:3.8-alpine
Unable to find image 'python:3.8-alpine' locally
3.8-alpine: Pulling from library/python
df20fa9351a1: Pull complete
36b3adc4ff6f: Pull complete
e638f3f9a1be: Pull complete
ab9516192caf: Pull complete
2d3173ad3f64: Pull complete
Digest: sha256:233f5f16c58103b614eff576ff8d6f8fe665ccc9399b49e8fc1f734ad3406cdf
Status: Downloaded newer image for python:3.8-alpine
Python 3.8.5 (default, Jul 21 2020, 18:15:54)
[GCC 9.3.0] on linux
Type "help", "copyright", "credits" or "license" for more information.
```

图 2-7　运行 Docker 镜像

```
Windows PowerShell
Windows PowerShell
Copyright (C) Microsoft Corporation. All rights reserved.

Try the new cross-platform PowerShell https://aka.ms/pscore6

PS C:\Users\yunp> docker run -ti python:3.8-alpine
Unable to find image 'python:3.8-alpine' locally
3.8-alpine: Pulling from library/python
df20fa9351a1: Pull complete
36b3adc4ff6f: Pull complete
e638f3f9a1be: Pull complete
ab9516192caf: Pull complete
2d3173ad3f64: Pull complete
Digest: sha256:233f5f16c58103b614eff576ff8d6f8fe665ccc9399b49e8fc1f734ad3406cdf
Status: Downloaded newer image for python:3.8-alpine
Python 3.8.5 (default, Jul 21 2020, 18:15:54)
[GCC 9.3.0] on linux
Type "help", "copyright", "credits" or "license" for more information.
>>> exit()
PS C:\Users\yunp> docker ps -a
CONTAINER ID   IMAGE               COMMAND     CREATED         STATUS                     PORTS   NAMES
59ebaca094b2   python:3.8-alpine   "python3"   4 minutes ago   Exited (0) 7 seconds ago           hardcore_keller
PS C:\Users\yunp>
```

图 2-8　列出处于运行状态的容器

如果要彻底清除该容器，则使用 docker rm 命令根据对应的容器 id 进行清除，如图 2-9 所示。

```
Windows PowerShell

Try the new cross-platform PowerShell https://aka.ms/pscore6

PS C:\Users\yunp> docker run -ti python:3.8-alpine
Unable to find image 'python:3.8-alpine' locally
3.8-alpine: Pulling from library/python
df20fa9351a1: Pull complete
36b3adc4ff6f: Pull complete
e638f3f9a1be: Pull complete
ab9516192caf: Pull complete
2d3173ad3f64: Pull complete
Digest: sha256:233f5f16c58103b614eff576ff8d6f8fe665ccc9399b49e8fc1f734ad3406cdf
Status: Downloaded newer image for python:3.8-alpine
Python 3.8.5 (default, Jul 21 2020, 18:15:54)
[GCC 9.3.0] on linux
Type "help", "copyright", "credits" or "license" for more information.
>>> exit()
PS C:\Users\yunp> docker ps -a
CONTAINER ID        IMAGE               COMMAND             CREATED             STATUS                      PORTS               NAMES
59ebaca094b2        python:3.8-alpine   "python3"           4 minutes ago       Exited (0) 7 seconds ago                        hardcore_keller
PS C:\Users\yunp> docker rm 59ebaca094b2
59ebaca094b2
PS C:\Users\yunp> docker ps -a
CONTAINER ID        IMAGE               COMMAND             CREATED             STATUS              PORTS               NAMES
PS C:\Users\yunp>
```

图 2-9 删除容器

如果要创建一个在退出后自动清理的容器，则在创建时需添加--rm 参数，如图 2-10 所示。

```
Windows PowerShell
PS C:\Users\yunp> docker run -ti --rm python:3.8-alpine
Python 3.8.5 (default, Jul 21 2020, 18:15:54)
[GCC 9.3.0] on linux
Type "help", "copyright", "credits" or "license" for more informatio
>>> exit()
PS C:\Users\yunp> docker ps -a
CONTAINER ID        IMAGE               COMMAND             CREATED
PS C:\Users\yunp>
```

图 2-10 启动一个停止时自动清除的容器

如果是服务器程序，需要长期运行，而且不需要可交互的终端，则可使用-d 参数运行，接下来我们以启动一个 apache 服务器为例来演示，如图 2-11 所示。

```
Windows PowerShell
PS C:\Users\yunp> docker run -d --rm httpd:alpine
Unable to find image 'httpd:alpine' locally
3625182e935a: Downloading  12.71MB/14.77MB
df20fa9351a1: Already exists
a036ef3df2d8: Pull complete
b2be8d8e0ee6: Pull complete
3625182e935a: Downloading  12.54MB/14.77MB
e8e86896757e: Download complete
3625182e935a: Downloading  13.2MB/14.77MB
3625182e935a: Pull complete
e8e86896757e: Pull complete
Digest: sha256:834d4ea86871e7ef849a62b7adcdd23580b8cf8b8b716ef0088da7ac2bdf2eb
5Status: Downloaded newer image for httpd:alpine
619d652ca28fe1b40b7d38f4503fe8193aa8208ccb2f783b31ead4af49870c0c
PS C:\Users\yunp>
```

图 2-11 启动可后台运行的容器

此时该容器处于运行状态，如图 2-12 所示。

虽然服务器处于运行状态，但是此时仍然无法访问，因为我们在启动时没有设置端口映射，现在我们使用 docker stop 命令来停止该容器，如图 2-13 所示。

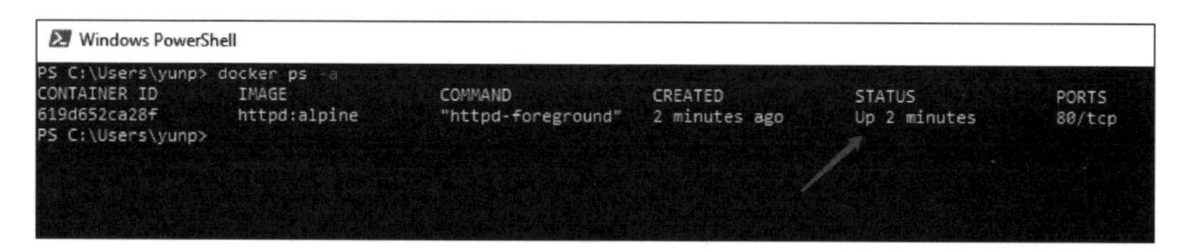

图 2-12 查看容器状态

```
Windows PowerShell
PS C:\Users\yunp> docker ps -a
CONTAINER ID    IMAGE          COMMAND              CREATED
619d652ca28f    httpd:alpine   "httpd-foreground"   2 minutes ago
PS C:\Users\yunp> docker stop 619d652ca28f
619d652ca28f
PS C:\Users\yunp> docker ps -a
CONTAINER ID    IMAGE          COMMAND              CREATED
PS C:\Users\yunp>
```

图 2-13 停止容器

现在我们重新启动该服务器，并使用-p 参数来配置端口映射，如图 2-14 所示。

```
Windows PowerShell
PS C:\Users\yunp> docker ps -a
CONTAINER ID    IMAGE          COMMAND              CREATED
619d652ca28f    httpd:alpine   "httpd-foreground"   2 minutes ago
PS C:\Users\yunp> docker stop 619d652ca28f
619d652ca28f
PS C:\Users\yunp> docker ps -a
CONTAINER ID    IMAGE          COMMAND              CREATED
PS C:\Users\yunp> docker run -d --rm -p 80:80 httpd:alpine
e43a22ca5650eada507be827e6ae7500bb4a9cfb0afe7a68d151ad1012f6f2c6
PS C:\Users\yunp>
```

图 2-14 Docker 端口映射

如果这是第一次运行该命令，将会弹出防火墙提示，此处需允许端口映射，如图 2-15 所示。

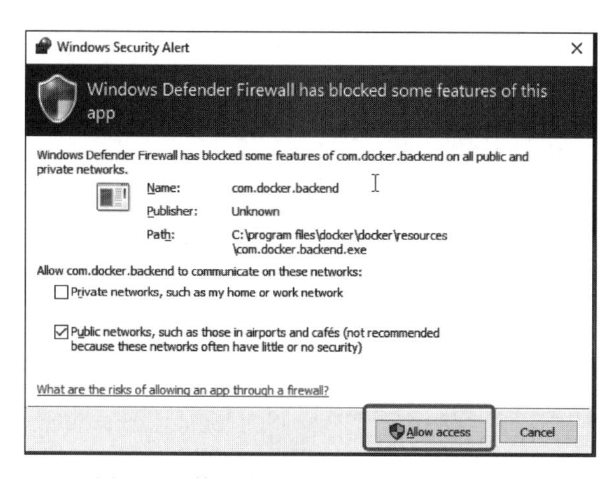

图 2-15 第一次运行服务器时防火墙提示

之后在浏览器地址栏中输入 http://127.0.0.1 以访问该站点，可以看到效果如图 2-16 所示。

图 2-16　服务器运行效果

5min

2.3　编写 Docker 镜像

我们已经学会了在终端使用 Docker 命令，但是如果仅仅如此，不但没有减少运维的工作量，反而增加了运维的难度和工作量，然而 Docker 的功能远不止这些，Docker 还支持编写镜像，可以把运行环境的配置工作脚本化，而后一键生成。

接下来以第 1 章/aio_http_server 项目中的 HTTP 服务器代码为例来编写一个 Docker 镜像。将第 1 章/aio_http_server 项目复制为第 2 章/aio_http_server，在项目根目录中创建一个 Dockerfile 文件，并输入如下代码：

```
#第 2 章/aio_http_server/Dockerfile

#指定该镜像基于 python:3.8 - alpine
FROM python:3.8 - alpine
#将 aio_http_server.py 文件复制到镜像中的 /opt 目录下
COPY aio_http_server.py /opt/
#指定该镜像启动后的默认运行命令，通过该命令启动 HTTP 服务器
CMD python /opt/aio_http_server.py
```

在终端执行命令 docker build . -t http_server 以构建该镜像，如图 2-17 所示。

```
Terminal:  Local ×  +
(venv) yunp@yunps-MBP http_server % docker build . -t http_server
Sending build context to Docker daemon  12.84MB
Step 1/3 : FROM python:3.8-alpine
 ---> f8a57363ff96
Step 2/3 : COPY server.py /opt/
 ---> 4ad679c078bd
Step 3/3 : CMD python /opt/server.py
 ---> Running in b50e4603dda1
Removing intermediate container b50e4603dda1
 ---> 62ceaf1268a5
Successfully built 62ceaf1268a5
Successfully tagged http_server:latest
(venv) yunp@yunps-MBP http_server %
```

图 2-17　构建镜像

接下来使用命令 docker images 以查看所有镜像，如图 2-18 所示。

```
Terminal: Local × +
(venv) yunp@yunps-MBP http_server % docker images
REPOSITORY          TAG                 IMAGE ID            CREATED             SIZE
http_server         latest              62ceaf1268a5        2 minutes ago       80.3MB
python              3.8-alpine          f8a57363ff96        38 hours ago        80.3MB
httpd               alpine              e7e8868c7697        2 weeks ago         56.1MB
```

图 2-18 查看镜像

可以看到出现了一个新的镜像名为 http_server，接下来运行该镜像，如图 2-19 所示。

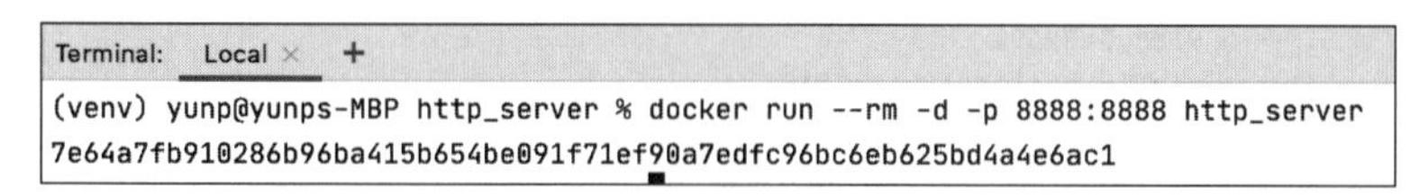

图 2-19 运行自定义镜像

这时服务器已经启动，可使用浏览器访问 http://127.0.0.1:8888，此时看到结果如图 2-20 所示。

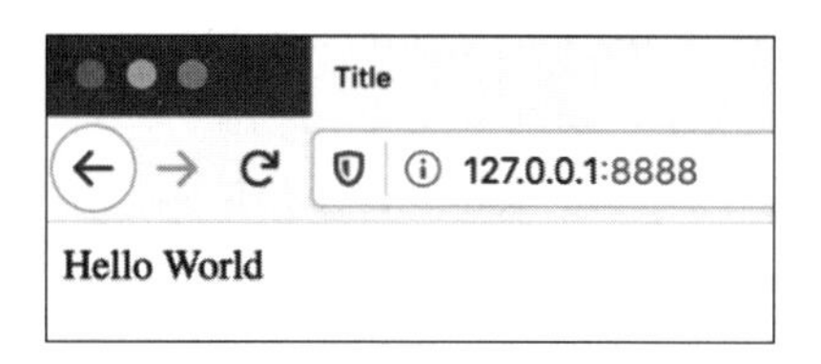

图 2-20 服务器访问结果

2.4 编排服务

虽然编写镜像已经可以省去大量的重复性工作，但是使用镜像还需要几个命令才能完成，尤其是涉及多服务的应用(如集成网站服务与数据库服务)，使用命令行的工作还是有些烦琐，这时 docker-compose 便派上用场了。

以第 2 章/aio_http_server 的 aio_http_server 为例进行说明，在项目的根目录下再创建一个 docker-compose.yml 文件，在其中输入代码如下：

```
#第 2 章/aio_http_server/docker-compose.yml
version: "3"

services:
  web:
    build: .
    ports:
      - "8888:8888"
```

该文件配置了一个名为 web 的服务器，使用当前目录进行镜像构建，并且将映射端口配置为 8888:8888，然后在终端只需执行命令 docker-compose up-d 便可自动构建，完成后便可运行，真正实现了一键启动，如图 2-21 所示。

```
Terminal:  Local ×  +
(venv) yunp@yunps-MBP http_server % docker-compose up -d
Creating network "http_server_default" with the default driver
Building web
Step 1/3 : FROM python:3.8-alpine
 ---> f8a57363ff96
Step 2/3 : COPY server.py /opt/
 ---> e478bdcbe452
Step 3/3 : CMD python /opt/server.py
 ---> Running in d0eb8591069c
Removing intermediate container d0eb8591069c
 ---> 93cba0c32691
Successfully built 93cba0c32691
Successfully tagged http_server_web:latest
WARNING: Image for service web was built because it did not already
Creating http_server_web_1 ... done _
```

图 2-21 实现一键启动

若要停止服务，则使用 docker-compose stop 命令。

若要停止并清除服务，则使用 docker-compose down 命令。

若要查看日志，则使用 docker-compose logs 命令。

若要查看服务状态，则使用 docker-compose ps 命令。

若要查看实时输出信息，则使用 docker-compose up 命令将服务运行进程切换到终端前台。

若要将处于终端前台的服务切换到后台，则直接使用快捷键 Ctrl+Z。

以上几个命令可以应对绝大多数需求，相当强大，所以 Docker 这工具只要使用过一次，便再也不能没有它。任何一个后端开发人员都应该学习使用 Docker。

第3章 AIOHTTP

AIOHTTP 是一个基于 asyncio 开发的 Python Web 开发框架，其稳定、强大、支持高并发。目前已实现的功能有 HTTP 服务器、HTTP 客户端、WebSocket 服务器，以及 WebSocket 客户端，服务功能还实现了路由、模板、会话(Session)等所有 Web 开发必备的功能模块。

3.1 创建异步 Web 服务器

5min

AIOHTTP 的使用非常简单，只需创建一个 aiohttp.web.Application 对象，并启动它便可完成一个 Web 服务器的最基本功能，代码如下：

```
from aiohttp import web

#创建一个服务器应用
app = web.Application()
#启动服务器应用
web.run_app(app)
```

启动前需要使用命令 pip install aiohttp 安装第三方依赖项 aiohttp。因为该服务器暂未编写任何页面请求处理逻辑，所以对其任何页面的访问都将出现 404 代码(找不到资源)，如图 3-1 所示。

图 3-1 空服务器示例

若要让其可以处理请求，则需要创建一个请求处理函数，该函数是一个异步函数，接收的一个传入参数是请求对象(aiohttp.web.Request)，返回一个响应对象(aiohttp.web.Respone)，并将该函数映射到服务器指定的路径上，代码如下：

```
"""
第 3 章/aiohttp_simple_server/server.py
```

```
"""
from aiohttp import web


#请求处理函数,接收一个参数为请求对象,返回信息将被传到浏览器端
async def hello(request: web.Request):
    return web.Response(text = "Hello, world")


#创建一个服务器应用
app = web.Application()
#将 hello 处理函数映射到网站根路径 /
app.router.add_get("/", hello)
#启动服务器应用
web.run_app(app)
```

在该示例中,将 hello 函数映射到网站根路径上,这时访问网站主页,将看到页面中显示信息 Hello, world,结果如图 3-2 所示。

图 3-2 AIOHTTP Hello,world 示例

接下来为该服务器配置容器化支持,在项目目录下创建 Dockerfile 文件,在其中输入如下代码:

```
#第 3 章/aiohttp_simple_server/Dockerfile
FROM python:3.8-slim
RUN pip install -i https://pypi.tuna.tsinghua.edu.cn/simple aiohttp
```

脚本指定该容器基于 python:3.8-slim,这是一个功能完整但轻量级的 Linux 环境,在第 2 行使用清华镜像安装项目依赖项 aiohttp。

在项目目录下创建一个 docker-compose.yml 文件,在其中输入如下代码:

```
#第 3 章/aiohttp_simple_server/docker-compose.yml
version: "3"

services:
  web:
    build: .
    ports:
      - "8080:8080"
    #如果指定该容器,则会在意外停止后自动重启
    restart: always
    #配置路径映射,将当前目录映射到容器中的 /opt 目录
    volumes:
      - ".:/opt/"
    #指定容器中应用程序的工作目录为 /opt
    working_dir: "/opt/"
    #配置容器启动时执行的命令,这里我们启动服务器
    command: python3 ./server.py
```

所有工作已经准备就绪，下面在终端里键入 docker-compose up-d 以自动构建镜像并启动，如图 3-3 所示。

```
Terminal:   Local ×   +
(venv) yunp@yunps-MacBook-Pro simple_aiohttp_server % docker-compose up -d
Creating network "simple_aiohttp_server_default" with the default driver
Creating simple_aiohttp_server_web_1 ... done
(venv) yunp@yunps-MacBook-Pro simple_aiohttp_server %
```

图 3-3　启动 Docker 服务

3.2　路由

4min

路由(route)是一种将路径映射到处理函数的机制，可以一个一个地单独配置，示例代码如下：

```
"""
第 3 章/aiohttp_route/config_route_one_by_one.py
"""
from aiohttp import web


async def hello(request: web.Request):
    return web.Response(text = "Hello, world")


app = web.Application()
app.router.add_get("/", hello)
web.run_app(app)
```

可以批量配置，代码如下：

```
"""
第 3 章/aiohttp_route/config_routes.py
"""
from aiohttp import web


async def hello(request: web.Request):
    return web.Response(text = "Hello, world")


async def users(req):
    return web.Response(text = "All users are here")


app = web.Application()
app.add_routes([
```

```
    web.get("/", hello),
    web.get("/users", users)
])
web.run_app(app)
```

也可以通过装饰器进行配置，代码如下：

```
"""
第3章/aiohttp_route/config_route_with_decorator.py
"""
from aiohttp import web

routes = web.RouteTableDef()


@routes.get('/')
async def hello(request):
    return web.Response(text = "Hello, world")


app = web.Application()
app.add_routes(routes)
web.run_app(app)
```

还可以通过路由配置接收其他的 HTTP 方法，代码如下：

```
@routes.post('/save_user_info')
async def save_user_info(req):
    return web.Response(text = "Saved")
```

如果仅仅支持配置路径映射，那么适用的场景就太少了，AIOHTTP 的路由还支持路径参数。例如，在实现一个博客类的网站时，通常会根据文章的 id 来定位该篇文章，为了对搜索引擎更友好，我们还会将文章的地址静态化，将类似 http://xxx.com? p=12 这样的路径改为 http://xxx.com/p/12，这就是服务器端开发常用的伪静态技术。代码如下：

```
"""
第3章/aiohttp_route/pages.py
"""
from aiohttp import web

routes = web.RouteTableDef()


@routes.get('/p/{page_id}')
async def page(req: web.Request):
    return web.Response(text = f"Page id is {req.match_info['page_id']}")


@routes.get('/')
```

```
async def hello(request):
    return web.Response(text = "Hello, world")


app = web.Application()
app.add_routes(routes)
web.run_app(app)
```

访问页面结果如图 3-4 所示。

图 3-4 解析路径参数

此外还可以使用正则表达式对路径中的参数进行约束，如果我们想让 page_id 只接收数字类型的参数，则可对代码进行修改，代码如下：

```
@routes.get(r'/p/{page_id:\d+}')
async def page(req: web.Request):
    return web.Response(text = f"Page id is {req.match_info['page_id']}")
```

本节中我们一共写了 4 个单独的示例文件，如图 3-5 所示。

接下来我们打算用一个 docker-compose 配置文件一键启动所有示例，并运行在不同的端口上，此时需要在项目目录中创建文件 Dockerfile 和 docker-compose.yml，文件结构如图 3-6 所示。

在 Dockerfile 中输入代码如下：

```
#第 3 章/aiohttp_route/Dockerfile
FROM python:3.8-slim
RUN pip install -i https://pypi.tuna.tsinghua.edu.cn/simple aiohttp
```

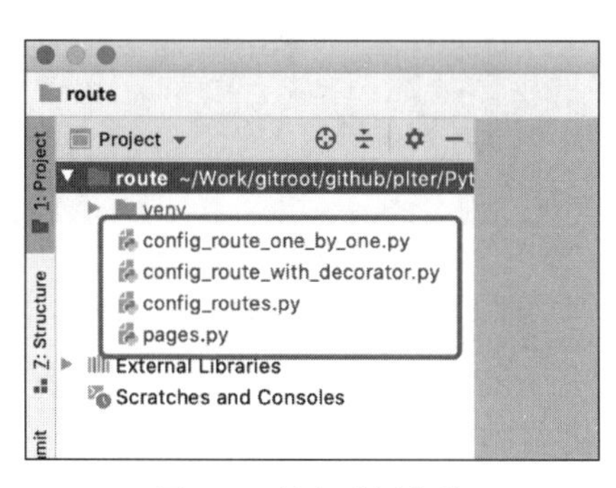

图 3-5 所有示例文件

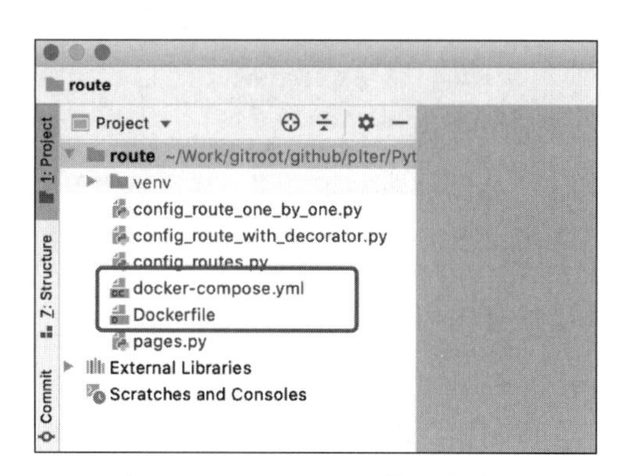

图 3-6 Docker 配置文件所在位置

在 docker-compose.yml 中输入代码如下：

```
#第 3 章/aiohttp_route/docker-compose.yml
version: "3"

services:
```

```
config_route_one_by_one:
  build: .
  ports:
    - "8080:8080"
  restart: always
  volumes:
    - ".:/opt/"
  working_dir: "/opt/"
  command: python3 ./config_route_one_by_one.py

config_route_with_decorator:
  build: .
  ports:
    - "8081:8080"
  restart: always
  volumes:
    - ".:/opt/"
  working_dir: "/opt/"
  command: python3 ./config_route_with_decorator.py

config_routes:
  build: .
  ports:
    - "8082:8080"
  restart: always
  volumes:
    - ".:/opt/"
  working_dir: "/opt/"
  command: python3 ./config_routes.py

pages:
  build: .
  ports:
    - "8083:8080"
  restart: always
  volumes:
    - ".:/opt/"
  working_dir: "/opt/"
  command: python3 ./pages.py
```

在该配置文件中创建了4个服务，将端口分别映射在本机的8080、8081、8082、8083端口，启动后可以分别通过这4个端口访问4个不同的示例。

3.3 静态文件处理

一个完整的网站除了服务器应用之外，一定会存在很多静态文件，例如图片、js脚本等。

在大型网站的部署上，静态资源会单独使用CDN进行管理，所以应用层可以不用考虑这个问题，直接通过CDN去引用静态资源文件即可。

在中小型网站的部署上,传统阻塞型IO编程模式下,由于应用服务器并不擅长处理高并发请求,一般来说会把应用服务器与基础HTTP服务器(如:Apache或者Nginx)集成并把静态文件的处理工作交给基础HTTP服务器。但在异步IO编程模型下,完全可以直接使用Python来处理静态资源请求。

AIOHTTP已经内置了静态资源文件处理的功能,因此可以直接使用,非常方便,代码如下:

```
"""
第3章/aiohttp_static_resource/server.py
"""

from aiohttp import web
import os

routes = web.RouteTableDef()


@routes.get('/')
async def hello(request):
    return web.Response(text = "Hello, world")


app = web.Application()
app.add_routes(routes)
#配置一个静态文件目录
app.router.add_static(
    "/static",
    os.path.join(os.path.dirname(__file__), 'static')
)
web.run_app(app)
```

配置一个静态文件目录,通过网站的/static路径进行访问,对应的文件都配置在项目目录下的static目录中,项目结构如图3-7所示。

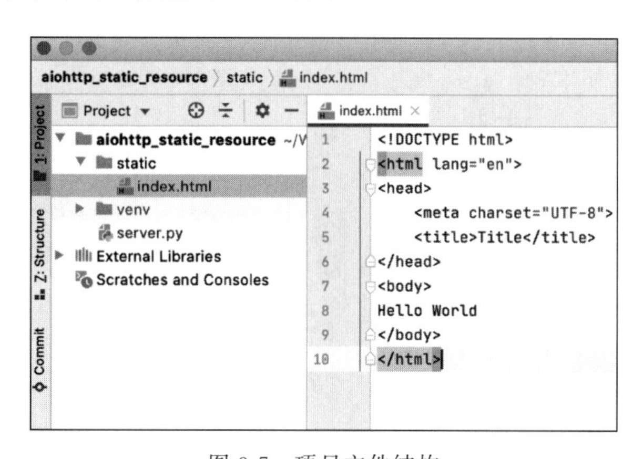

图3-7 项目文件结构

启动服务器后可通过浏览器访问地址 http://127.0.0.1:8080/static/index.html，结果如图 3-8 所示。

图 3-8　静态文件访问结果

与该项目相关的 Dockerfile 文件内容如下：

```
#第 3 章/aiohttp_static_resource/Dockerfile
FROM python:3.8-slim
RUN pip install -i https://pypi.tuna.tsinghua.edu.cn/simple aiohttp
```

与该项目相关的 docker-compose.yml 文件内容如下：

```
#第 3 章/aiohttp_static_resource/docker-compose.yml
version: "3"

services:
  web:
    build: .
    ports:
      - "8080:8080"
    restart: always
    volumes:
      - ".:/opt/"
    working_dir: "/opt/"
    command: python3 ./server.py
```

9min

3.4　模板渲染

如果每个页面都只是直接使用字符串拼接而成，那么实际的工作量也是很大的，所以人们发明了模板技术，用现成的 HTML 文件经过处理（渲染），并将处理之后的结果返回给浏览器端就可以大大节省开发成本。

在 AIOHTTP 服务端应用开发时，可配套使用 aiohttp_jinja2 库实现模板渲染，通过命令 pip install aiohttp_jinja2 进行安装。首先需要指定模板文件的根目录，并将模板系统与服务端应用集成，代码如下：

```
aiohttp_jinja2.setup(
    app,
    loader=jinja2.FileSystemLoader(
        os.path.join(os.path.dirname(__file__), "tpls")
    )
)
```

然后在需要渲染的请求处理函数前添加装饰器以指定要渲染的模板的路径，代码如下：

```
@routes.get('/')
@aiohttp_jinja2.template("index.html")
```

```
async def hello(request):
    return dict()
```

完整的代码如下：

```
"""
第 3 章/aiohttp_template/server.py
"""

from aiohttp import web
import jinja2
import aiohttp_jinja2
import os

app = web.Application()
aiohttp_jinja2.setup(
    app,
    loader = jinja2.FileSystemLoader(
        os.path.join(os.path.dirname(__file__), "tpls")
    )
)
routes = web.RouteTableDef()


@routes.get('/')
@aiohttp_jinja2.template("index.html")
async def hello(request):
    return dict()


app.add_routes(routes)
web.run_app(app)
```

项目文件结构如图 3-9 所示。

其中模板文件 index.html 的内容如下：

```
<!-- 第 3 章/aiohttp_template/tpls/index.html -->
<!DOCTYPE html>
<html lang = "en">
<head>
<meta charset = "UTF-8">
<title>Title</title>
</head>
<body>
Hello World
</body>
</html>
```

启动服务器后,最终浏览器运行结果如图 3-10 所示。

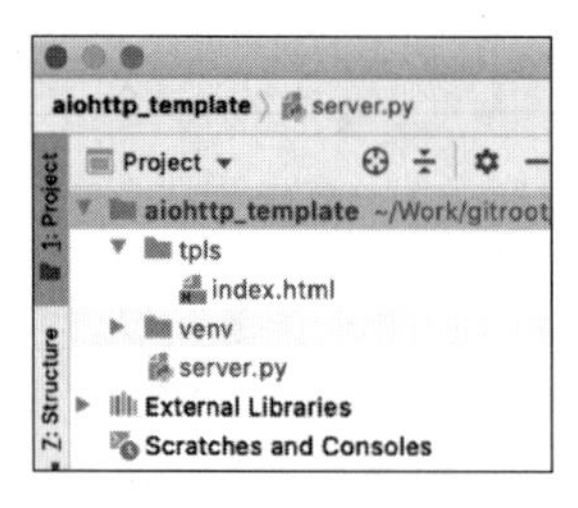

图 3-9 项目文件结构

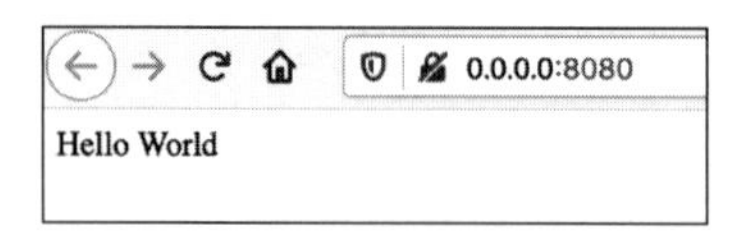

图 3-10 模板渲染结果

如果仅仅读取静态文件并且呈现出来,那么模板技术显得没什么用处。而实际上,模板技术能够允许我们从后端传参数给模板并最终由模板引擎渲染成页面,例如从数据库中读取用户的信息并传给模板。

若要从 Python 代码中向模板文件传参数,只需请求处理函数返回一个字典,代码如下:

```
@routes.get('/')
@aiohttp_jinja2.template("index.html")
async def hello(request):
    return dict(name = "小云", age = 20)
```

模板文件中使用参数的代码如下:

```
<!-- 第 3 章/aiohttp_template/tpls/index.html -->
<!DOCTYPE html>
<html lang = "en">
<head>
<meta charset = "UTF - 8">
<title>Title</title>
</head>
<body>
<ul>
<li>名字:{{ name }}</li>
<li>年龄:{{ age }}</li>
</ul>
</body>
</html>
```

最终页面渲染的结果如图 3-11 所示。

此外,为了减少不必要的重复性工作,模板还支持继承、循环、条件渲染等。下面我们以常规网站页面为例,深入了解更多与模板引擎使用相关的知识。

实现效果如图 3-12 所示。

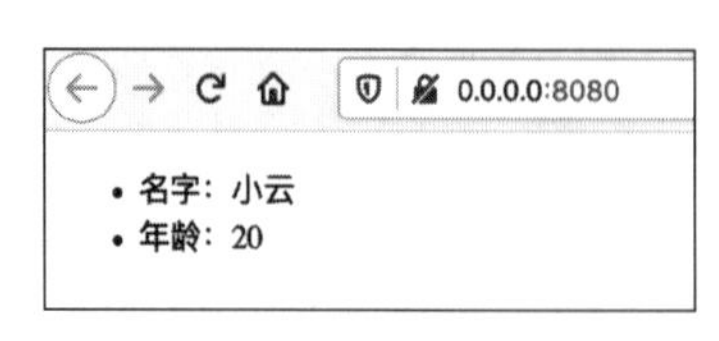

图 3-11 向模板传参数

图 3-12 通用网站框架结构

主页按钮背景和页面标题会随着页面的变化而变化，在博客页面还渲染了一个列表，如图 3-13 所示。

图 3-13 博客页面示例

在该项目的实现过程中，还使用了前端界面框架 Bootstrap，接下来一步一步实现这个常规网站的基本框架。

使用 PyCharm 创建一个新的项目，并命名为 aiohttp_template_pages，在终端中使用 npm init 命令初始化 npm 环境（使用该命令之前需要先安装 nodejs，到 https://nodejs.org 下载并安装）。所有选项均选择默认即可（在所有遇到需要输入信息时单击回车键直到完成），如图 3-14 所示。

```
Terminal:  Local ×  +
(venv) yunp@yunps-MBP aiohttp_template_pages % npm init
This utility will walk you through creating a package.json file.
It only covers the most common items, and tries to guess sensible defaults.

See `npm help init` for definitive documentation on these fields
and exactly what they do.

Use `npm install <pkg>` afterwards to install a package and
save it as a dependency in the package.json file.

Press ^C at any time to quit.
package name: (aiohttp_template_pages)
version: (1.0.0)
description:
entry point: (index.js)
test command:
git repository:
keywords:
author:
license: (ISC)
About to write to /Users/yunp/Work/gitroot/github/plter/PythonAIOProgramming/

{
  "name": "aiohttp_template_pages",
  "version": "1.0.0",
  "description": "",
  "main": "index.js",
  "scripts": {
    "test": "echo \"Error: no test specified\" && exit 1"
  },
  "author": "",
  "license": "ISC"
}


Is this OK? (yes)
(venv) yunp@yunps-MBP aiohttp_template_pages %
```

图 3-14 初始化 npm 项目

命令执行结束后将在项目目录下创建一个 package.json 文件，如图 3-15 所示。

通过 npm 命令 npm install--save jquery bootstrap--registry=https://registry.npm.taobao.org 安装前端依赖项 jQuery 和 Bootstrap，在执行该命令时可指定使用淘宝的仓库，这样速度更快，执行完毕后，将在 node_modules 目录下新增 bootstrap 和 jquery 两个目录，如图 3-16 所示。

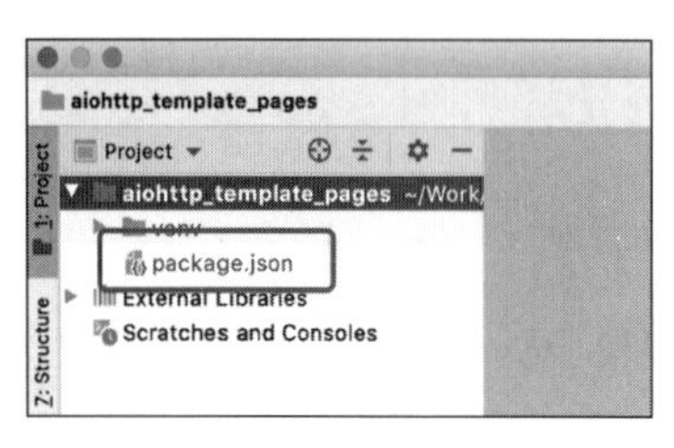

图 3-15 package.json 文件

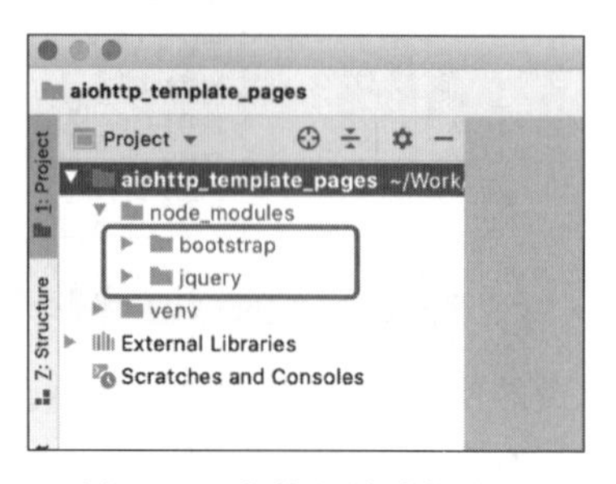

图 3-16 安装的前端依赖项

建立 tpls 目录，用于存放模板文件，在其中创建 layout.html、index.html、news.html、blog.html 4 个文件，结构如图 3-17 所示。

其中 blog.html、index.html、news.html 3 个模板分别对应 3 个页面，这 3 个页面的框架布局是一样的，所以我们将框架抽象出来并创建一个父级模板 layout.html。

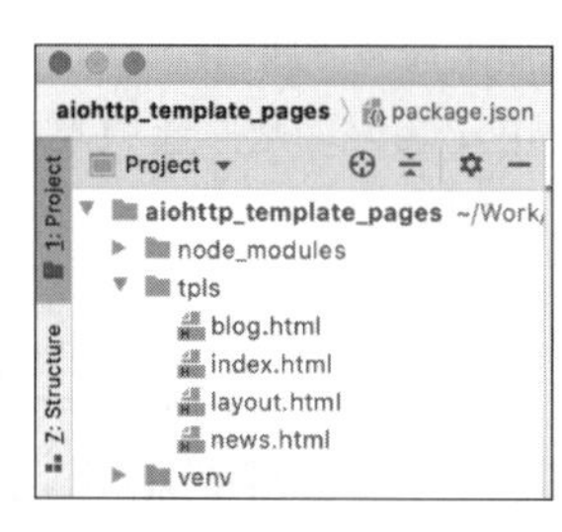

图 3-17 模板文件

其中模板 layout.html 文件的代码如下：

```
<!-- 第 3 章/aiohttp_template_pages/tpls/layout.html -->
<!DOCTYPE html>
<html lang="en">
<head>
<meta charset="UTF-8">
<!--
要求传一个名为 title 的参数,这样不同的页面传来不同
的值以实现不同页面拥有不同标题的效果
    -->
<title>
        {{ title }}
</title>

<!--
引用前端页面的依赖项,使用的前端界面框架是 Bootstrap,
因为 Bootstrap 依赖 jQuery,所以我们也引用了 jQuery,
并且 jQuery 的引用语句必须写在 Bootstrap 引用语句的前面
     -->
<link rel="stylesheet" href="/node_modules/bootstrap/dist/css/bootstrap.min.css">
<script src="/node_modules/jquery/dist/jquery.min.js"></script>
<script src="/node_modules/bootstrap/dist/js/bootstrap.bundle.min.js"></script>
```

```
</head>
<body>
<nav class = "navbar navbar - light bg - light" style = "justify - content:start">
<!-- 配置顶部菜单 -->
<a class = "navbar - brand" href = "/">小云站</a>

<ul class = "nav nav - pills">
<li class = "nav - item">
<!--
使用了模板引擎支持的判断语句,根据当前页面路径选择性
地给菜单项添加 active 类,以使它处于激活状态,这里
需要注意的是我们使用了 request 这个参数,意味着在渲
染时需要传入 request 对象
            -->
<a class = "nav - link {% if request.path == '/' %}active{% endif %}"
            href = "/">
首页
</a>
</li>
<li class = "nav - item">
<a class = "nav - link {% if request.path == '/news' %}active{% endif %}"
            href = "/news">
动态
</a>
</li>
<li class = "nav - item">
<a class = "nav - link {% if request.path == '/blog' %}active{% endif %}"
            href = "/blog">
博客
</a>
</li>
</ul>
</nav>
<div class = "container - fluid">
<!-- 声明一个名为 body 的代码块,子模板可以重写该
代码块以自定义页面的内容 -->
    {% block body %}
    {% endblock %}
</div>
</body>
</html>
```

模板 index.html 的代码如下：

```
<!-- 第 3 章/aiohttp_template_pages/tpls/index.html -->
<!-- 指定继承模板 layout.html -->
{% extends "layout.html" %}
{% block body %}
```

```
<!-- 重写代码块 body 的内容 -->
这是首页
{% endblock %}
```

接下来创建 server.py 文件，用于配置服务器，其代码如下：

```
"""
第 3 章/aiohttp_template_pages/server.py
"""
from aiohttp import web
import jinja2
import aiohttp_jinja2
import os

app = web.Application()
aiohttp_jinja2.setup(
    app,
    loader=jinja2.FileSystemLoader(
        os.path.join(os.path.dirname(__file__), "tpls")
    )
)
routes = web.RouteTableDef()


@routes.get('/')
@aiohttp_jinja2.template("index.html")
async def index(request: web.Request):
    return dict(title="首页", request=request)


@routes.get('/news')
@aiohttp_jinja2.template("news.html")
async def news(request: web.Request):
    return dict(title="动态", request=request)


@routes.get('/blog')
@aiohttp_jinja2.template("blog.html")
async def blog(request: web.Request):
    return dict(
        title="博客", request=request,
        # 向模板传递数组
        posts=[
            {"title": "文章 1", "content": "内容 1"},
            {"title": "文章 2", "content": "内容 2"}
        ]
    )
```

```
app.add_routes(routes)
app.router.add_static(
    "/node_modules",
    os.path.join(os.path.dirname(__file__), "node_modules")
)
web.run_app(app)
```

其中/blog 页面向模板传递了数组，则对应的模板代码如下：

```
<!-- 第 3 章/aiohttp_template_pages/tpls/blog.html -->
{% extends "layout.html" %}
{% block body %}
这是博客页面
<div>
    {% for p in posts %}
<!-- 使用模板引擎支持的循环语句生成 html 代码 -->
<div>
<div class="font-weight-bold">{{ p.title }}</div>
<div>{{ p.content }}</div>
</div>
    {% endfor %}
</div>
{% endblock %}
```

与该项目对应的 Dockerfile 文件代码如下：

```
#第 3 章/aiohttp_template_pages/Dockerfile
FROM python:3.8-slim
RUN pip install -i https://pypi.tuna.tsinghua.edu.cn/simple aiohttp
RUN pip install -i https://pypi.tuna.tsinghua.edu.cn/simple aiohttp_jinja2
```

与该项目对应的 docker-compose.yml 代码如下：

```
#第 3 章/aiohttp_template_pages/docker-compose.yml
version: "3"

services:
  web:
    build: .
    ports:
      - "8080:8080"
    restart: always
    volumes:
      - ".:/opt/"
    working_dir: "/opt/"
    command: python3 ./server.py
```

6min

3.5 处理表单提交

处理表单提交是一个 Web 服务器最基本的功能，AIOHTTP 也提供了强大的表单处理功能，首先看一下效果，如图 3-18 所示。

图 3-18　表单页面渲染效果

在表单中输入信息后提交则可呈现结果页面，如图 3-19 所示。

图 3-19　表单提交后呈现结果

这个项目使用了 Bootstrap 来搭建前端页面，同样的使用流程不再赘述，这里只讲解最核心的处理逻辑。

在首页中编写一个表单，代码如下：

```
<!-- 第 3 章/aiohttp_submitform/tpls/index.html -->
{% extends "layout.html" %}
{% block body %}
<!-- 配置该表单提交的方式为 post 方式，提交的目标页面路径
是/login，并且以 URL 参数对的形式提交数据 -->
<form method="post" action="/login" enctype="application/x-www-form-URLencoded">
<div class="card" style="margin: 2rem auto 0 auto;width:300px;">
<div class="card-header">登录</div>
```

```
<div class="card-body">
<div class="form-group row">
<label for="inputName" class="col-sm-3 col-form-label">姓名</label>
<div class="col-sm-9">
<!-- 指定姓名输入框的字段名为 name -->
<input type="text" name="name" class="form-control" id="inputName">
</div>
</div>
<div class="form-group row">
<label for="inputAge" class="col-sm-3 col-form-label">年龄</label>
<div class="col-sm-9">
<!-- 指定年龄输入框的字段名为 age -->
<input type="number" name="age" class="form-control" id="inputAge">
</div>
</div>
<div class="form-group row">
<div class="col-sm">
<button type="submit" class="btn btn-primary">登录</button>
</div>
</div>
</div>
</div>
</form>
{% endblock %}
```

该模板所继承的父级模板源码如下：

```
<!-- 第3章/aiohttp_submitform/tpls/layout.html -->
<!DOCTYPE html>
<html lang="en">
<head>
<meta charset="UTF-8">
<title>{{ title }}</title>
<link rel="stylesheet" href="/node_modules/bootstrap/dist/css/bootstrap.min.css">
<script src="/node_modules/jquery/dist/jquery.min.js"></script>
<script src="/node_modules/bootstrap/dist/js/bootstrap.bundle.min.js"></script>
</head>
<body>
{% block body %}
{% endblock %}
</body>
</html>
```

与/login 页面对应的处理函数源码如下：

```
@routes.post("/login")
@aiohttp_jinja2.template("login.html")
async def login(request: web.Request):
    # 读取表单数据
    user = await request.post()
    return dict(title="登录结果", user=user)
```

通过 request.post 函数解析前端以 post 方式传来的数据，并将数据传给模板，由模板进行渲染，与 /login 对应的模板文件源码如下：

```
<!-- 第 3 章/aiohttp_submitform/tpls/login.html -->
{ % extends "layout.html" % }
{ % block body % }
<div class = "card" style = "margin: 2rem auto 0 auto;width:300px;">
<div class = "card - header">接收的数据</div>
<div>
<table class = "table table - bordered">
<tbody>
<tr>
<td>姓名</td>
<td style = "width:200px;">{{ user.name }}</td>
</tr>
<tr>
<td>年龄</td>
<td>{{ user.age }}</td>
</tr>
</tbody>
</table>
</div>
</div>
{ % endblock % }
```

3.6 文件上传

6min

本节计划实现一个上传图片后将此图片呈现在页面中的过程。

上传表单如图 3-20 所示。

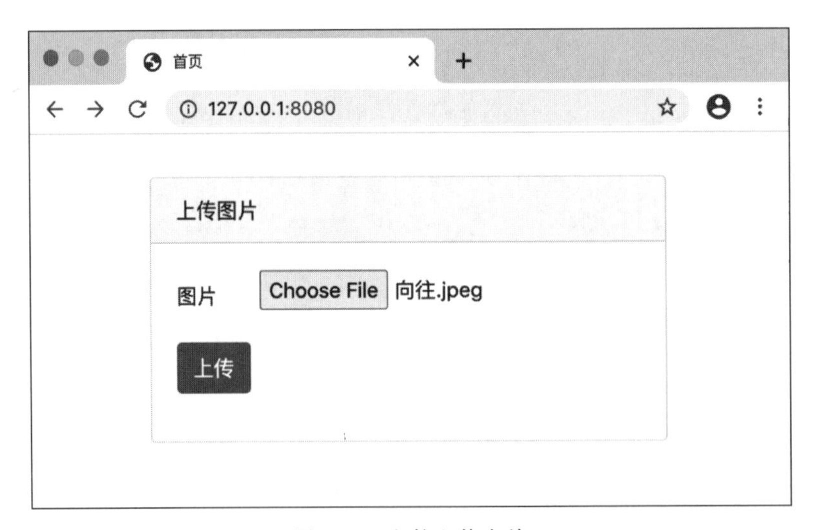

图 3-20　文件上传表单

选择一张图片，上传后效果如图 3-21 所示。

图 3-21 文件上传后的效果

上传文件表单的源码如下：

```
<!-- 第 3 章/aiohttp_upload_file/tpls/index.html -->
{% extends "layout.html" %}
{% block body %}
<!-- 指定提交表单的方式为 post,提交的目标页面是 /upload,
并且编码方式为 multipart/form-data,上传文件时,这几项
必须这样配置,否则后端将无法正确接收数据 -->
<form method="post" action="/upload" enctype="multipart/form-data">
<div class="card" style="margin: 2rem auto 0 auto;width:400px;">
<div class="card-header">上传图片</div>
<div class="card-body">
<div class="form-group row">
<label for="inputName" class="col-sm-2 col-form-label">图片</label>
<div class="col-sm-10">
<!-- 指定输入框的类型为 file,这会在
页面中呈现一个浏览本地文件的按钮 -->
<input type="file" name="file" id="inputName">
</div>
</div>
<div class="form-group row">
<div class="col-sm">
<button type="submit" class="btn btn-primary">上传</button>
</div>
</div>
</div>
</div>
</form>
{% endblock %}
```

则对应的服务端处理逻辑源码如下：

```
"""
第 3 章/aiohttp_upload_file/server.py
"""
from aiohttp import web
import os, datetime, aiofile, aiohttp_jinja2, jinja2

APP_ROOT = os.path.dirname(__file__)
app = web.Application()
aiohttp_jinja2.setup(
    app,
    loader = jinja2.FileSystemLoader(
        os.path.join(os.path.dirname(__file__), "tpls")
    )
)
routes = web.RouteTableDef()


@routes.get('/')
@aiohttp_jinja2.template("index.html")
async def index(request: web.Request):
    return dict(title = "首页")


@routes.post("/upload")
@aiohttp_jinja2.template("upload.html")
async def login(request: web.Request):
    #解析前端上传的数据
    data = await request.post()
    #根据表单字段名从上传数据中取得文件对象,与<input name = "file">对应
    file_object = data['file']
    #用当前时间当作要保存的文件名
    file_name = f"{datetime.datetime.now().timestamp()}"
    #以写入二进制数据(wb)的方式打开文件
    file = await aiofile.open_async(
        os.path.join(APP_ROOT, "uploads", file_name), "wb"
    )
    #将数据读取出来并保存到目标文件中
    await file.write(file_object.file.read())
    #关闭文件 IO
    await file.close()
    #将该文件在网站中的路径传给模板
    return dict(title = "上传结果", file_path = f"/uploads/{file_name}")


app.add_routes(routes)
app.router.add_static("/node_modules", os.path.join(APP_ROOT, "node_modules"))
app.router.add_static("/uploads", os.path.join(APP_ROOT, "uploads"))
web.run_app(app)
```

实现的基本逻辑是先解析上传的数据，从数据库获得文件对象，并将文件对象的内容读取出来，再以写入二进制数据的方式保存到指定的文件中，最后将保存后的文件在网站中的路径传给模板，由模板渲染出来。这里我们使用当前时间对文件进行重命名，目的是保证上传的文件名不重复，在实际使用中，这种方式并不是最可靠的，在高并发的情况下，有可能存在同一时间上传多个文件，所以在实际开发工作中，为了保证文件名的唯一性，我们可以把时间与用户 id 关联起来形成真正唯一的文件名。保证文件名唯一的方式还有很多种，例如使用 uuid，不管使用哪种方式，我们的目标是保证文件名唯一。

在这段代码里用到了 aiofile，这是在 1.5 节文件异步 IO 中所实现的库，如果还有印象可重复利用该代码，如果没有印象需要回去再看看那一节。

3.7 Session

会话(Session)是一种在服务器端缓存用户数据的机制，主要用于保存用户的登录信息。如果要在 AIOHTTP 中支持 Session，需要第三方库 aiohttp_session，用命令 pip install aiohttp_session 进行安装。

在使用 Session 功能之前需要先为应用配置 Session，代码如下：

```
aiohttp_session.setup(app, aiohttp_session.SimpleCookieStorage())
```

然后在请求处理函数中使用 Session 的代码如下：

```
session = await aiohttp_session.get_session(request)
```

可运行的完整示例代码如下：

```
"""
第 3 章/aiohttp_session_counter/server.py
"""
from aiohttp import web
import aiohttp_session

routes = web.RouteTableDef()


@routes.get('/')
async def index(request: web.Request):
    session = await aiohttp_session.get_session(request)
    session['count'] = (session['count'] if 'count' in session else 0) + 1
    return web.Response(text = f"Count is {session['count']}")


app = web.Application()
app.add_routes(routes)
aiohttp_session.setup(app, aiohttp_session.SimpleCookieStorage())
web.run_app(app)
```

这段程序实现了当前用户的访问计数，用户每刷新一次页面，则计数加1，如图3-22所示。

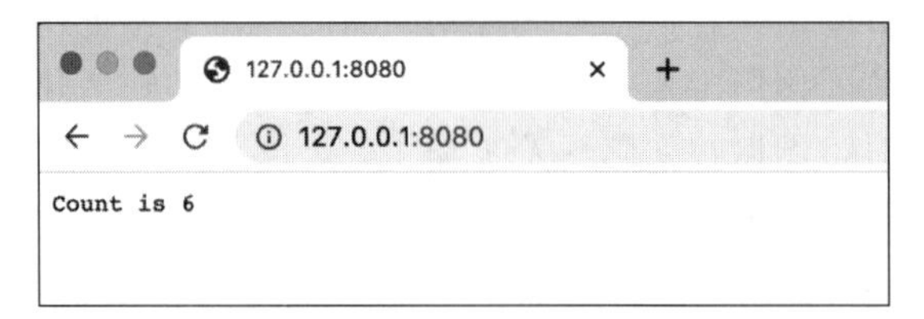

图3-22 Session计数器

如果希望以加密的方式在浏览器与服务器之间传输Session ID，则可使用aiohttp_session.cookie_storage.EncryptedCookieStorage进行实现，代码如下：

```
"""
第3章/aiohttp_ssession_counter/server.py
"""
import base64

from aiohttp import web
from aiohttp_session import setup, get_session
from aiohttp_session.cookie_storage import EncryptedCookieStorage
from cryptography import fernet

routes = web.RouteTableDef()


@routes.get('/')
async def index(request: web.Request):
    session = await get_session(request)
    session['count'] = \
        (session['count'] if 'count' in session else 0) + 1
    return web.Response(text = f"Count is {session['count']}")


app = web.Application()
app.add_routes(routes)
fernet_key = fernet.Fernet.generate_key()
#secret_key 必须是 URL 安全的、base64 编码的 32 字节数据
#URL 安全是指不能包括 '+' 与 '/', 在实际使用中会使用 '-' 代替 '+', 使用'_' 代替 '/'
secret_key = base64.URLsafe_b64decode(fernet_key)
setup(app, EncryptedCookieStorage(secret_key))
web.run_app(app)
```

与该示例配套的Dockerfile内容如下：

```
#第3章/aiohttp_ssession_counter/Dockerfile
FROM python:3.8-slim
RUN pip install -i https://pypi.tuna.tsinghua.edu.cn/simple \
    aiohttp cryptography aiohttp_session
```

配套的 docker-compose.yml 文件内容如下：

```
#第 3 章/aiohttp_ssession_counter/docker-compose.yml
version: "3"

services:
  web:
    build: .
    ports:
      - "8080:8080"
    restart: always
    volumes:
      - ".:/opt/"
    working_dir: "/opt/"
    command: python3 ./server.py
```

3.8 HTTP 客户端

AIOHTTP 库实现了完整而健壮的 HTTP 客户端。在 1.8 节异步 HTTP 客户端中，我们已经介绍过如何实现 HTTP 客户端，但是 HTTP 的标准非常多，如果按 HTTP 协议去逐个实现，工作量也是非常庞大的。但如果使用 AIOHTTP 来作为开发框架，则可以直接使用已经实现的、功能完整的 HTTP 客户端。

在不使用 AIOHTTP 作为 Web 开发框架的情况下，是否也可以使用 AIOHTTP 的客户端库呢？当然可以，AIOHTTP 的客户端库是一套功能完整的、独立实现的库，可以应用在任何 Python 异步编程中。

接下来使用 AIOHTTP 的客户端去连接第 3 章/aiohttp_session_counter 项目，客户端代码如下：

```
session = aiohttp.ClientSession() #创建客户端会话上下文
response = await session.get('http://127.0.0.1:8080') #连接服务器
print(f"[{time.strftime('%X')}] {await response.text()}") #读取返回数据
await session.close() #关闭会话
```

AIOHTTP 的客户端还可以模拟 Cookie 机制，这意味着 AIOHTTP 客户端可以很方便地用于模拟登录等其他复杂的应用场景，下面代码演示了 AIOHTTP 客户端对服务器端会话(Session)机制的支持：

```
"""
第 3 章/aiohttp_client/app.py
"""
import asyncio, time

import aiohttp
```

```
async def main():
    session = aiohttp.ClientSession(
        #使用最简单的 Cookie 机制
        cookie_jar = aiohttp.CookieJar(unsafe = True)
    )
    for i in range(10):
        response = await session.get('http://127.0.0.1:8080')
        print(f"[{time.strftime('%X')}] {await response.text()}")
        #每次请求后休眠 1s
        await asyncio.sleep(1)
    await session.close()


if __name__ == '__main__':
    asyncio.run(main())
```

代码运行的结果如图 3-23 所示。

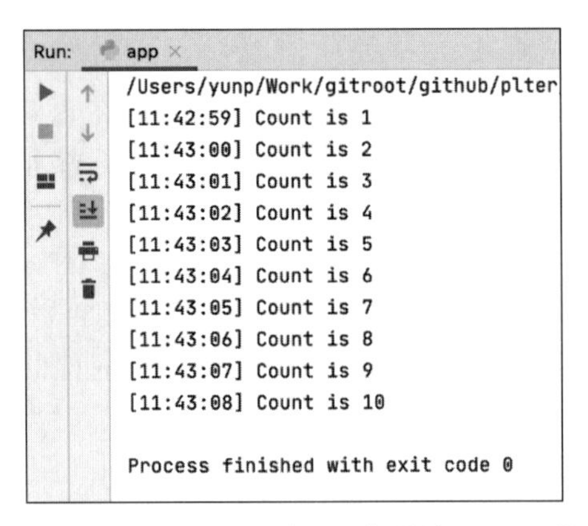

图 3-23 AIOHTTP 客户端对服务器端 Session 的支持

在使用 AIOHTTP 客户端时，只需创建一个 ClientSession 实例，之后可以使用该实例去执行所有的请求任务，强烈推荐这种方法，如果是同时请求多个网站，则需要分别为每个网站创建单独的 ClientSession 实例。

3.9 HTTPS 支持

互联网技术已经发展了很多年，至今在世界范围内有无数的网站，今天人类的生活已经很难完全脱离互联网。在我们享受科技的同时，也担心隐私泄露等问题，作为一家网站，有义务保护用户隐私，所以今天的世界要求每一家网站都通过 HTTPS 来服务用户。

每一家云主机服务商都提供免费的 HTTPS 证书，通常有效期为一年，到期后网站维护人员应重新申请证书并进行部署。所以将网站 HTTPS 化并不会增加成本，作为网站的运营者和开发者，更加有义务为用户提供安全的服务。

如果网站以 HTTP 直接部署，则用户在访问时浏览器会提示用户这是不安全的链接，如图 3-24 所示。

用户在访问这类网站时就会明白，隐私信息可能会泄露给第三方（例如木马、监控软件等），遇到这种情况用户一般会直接离开该网站，对于网站来说，也就会遇到极大的推广障碍。如果采用的是 HTTPS 的安全链接，则浏览器会提示安全或者不提示，如图 3-25 所示。

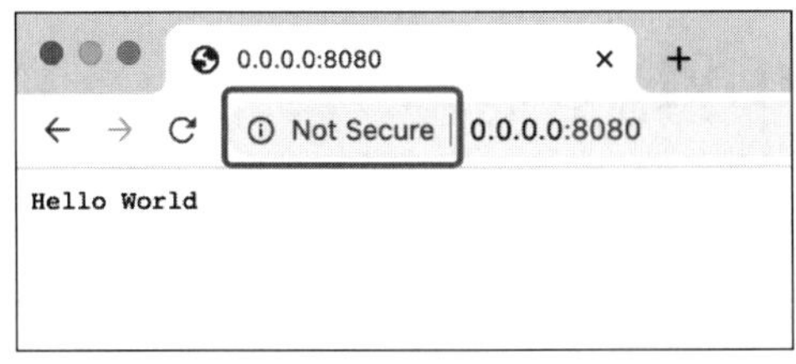

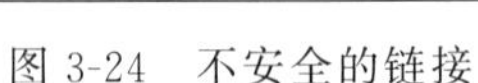
图 3-24　不安全的链接

图 3-25　安全链接

在 AIOHTTP 框架中，配置支持 HTTPS 非常方便，示例代码如下：

```
"""
第 3 章/aiohttps/server.py
"""
from aiohttp import web
import ssl, os

routes = web.RouteTableDef()


@routes.get('/')
async def index(request: web.Request):
    return web.Response(text = "Hello World")


app = web.Application()
app.add_routes(routes)
ssl_context = ssl.create_default_context(ssl.Purpose.CLIENT_AUTH)
#加载网站证书文件
ssl_context.load_cert_chain(
    os.path.join(os.path.dirname(__file__), "ssl", "cert.pem"),
    os.path.join(os.path.dirname(__file__), "ssl", "cert.key"),
)
web.run_app(app, port = 443, ssl_context = ssl_context)
```

在该示例中，使用两个文件，一个是证书文件，另一个是密钥文件，在你申请 HTTPS 证书完成后可以获得这两个文件。

AIOHTTP 创建的 HTTPS 服务器默认使用端口 8443，而 HTTPS 的默认端口号是 443，所以如果想使用 HTTPS 默认端口（访问时域名后可省略端口号），须手动指定使用 443 端口。

第 4 章

aioMySQL

aioMySQL 是一个基于 asyncio 的用于访问 MySQL/MariaDB 数据库的库，它依赖 PyMySQL 库，并且重用了 PyMySQL 的大部分代码。

事实上，使用 asyncio 框架去操作数据库的库不仅只有 aioMySQL，为了支持不同类型的数据，aiohttp 开发团队实现了多个用于操作不同数据库的库，如 aiopg 用于操作 PostgreSQL 数据库，aioodbc 用于操作 SQL Server 数据库。所有这些都是开源的，源码可在 https://GitHub.com/aio-libs 上找到。

4.1 搭建 MariaDB 数据库环境

4min

在开始之前，先介绍一下 MySQL 与 MariaDB 的关系。MySQL 是绝大多数开发者非常熟知的关系型数据库系统，但是提起 MariaDB，估计很多人还比较陌生，其实 MariaDB 的开发团队成员都来自 MySQL 的原始开发团队。

MySQL 原始开发团队秉承开源信仰，创造了世界上使用最广泛的开源数据库之一 MySQL，后来被 Sun 公司收购，虽然 Sun 是一家商业公司，但其领导层也热衷于支持开源事业，所以对 MySQL 的开源继续提供大力支持。

但是最终改变 MySQL 的事件是 Sun 公司被 Oracle 公司收购，如今的 MySQL 理所当然属于 Oracle 公司，而 Oracle 是一家纯粹的封闭式商业公司，而 MySQL 的开源信仰在 Oracle 看来就是个笑话。如今的 MySQL 已经闭源并且收费，当你打开 MySQL 的官方网站时，可以看到购买方式说明。

在这种情况下，MySQL 的原始开发团队成员组建了新的开发团队，基于最后的开源版 MySQL 5 继续开发、保持开源免费，这就是现在的 MariaDB。在赞美 MySQL 的原始开发团队时，也不能贬低 Oracle，作为一家商业公司，最重要的是活下去，不管有多大的梦想，如果活不下去，那都是空话，所以 Oracle 对 MySQL 用户收费是可以理解的。与此同时，MySQL 原始开发团队的成员，那些计算机科学巨匠们，继续为最初的信仰而努力着，这更是非常值得尊敬的。如果可以，请为那些你受惠过的开源项目提供力所能及的赞助，这是开源精神的火种。

搭建 MariaDB 环境有很多种方式，我们选择最容易操作的方式，也是最具有实战意义的方式，使用 Docker 来搭建 MariaDB 数据库环境。

创建一个新的 Python 项目，在项目的根目录下创建一个 docker-compose. yml 文件，在其中输入代码如下：

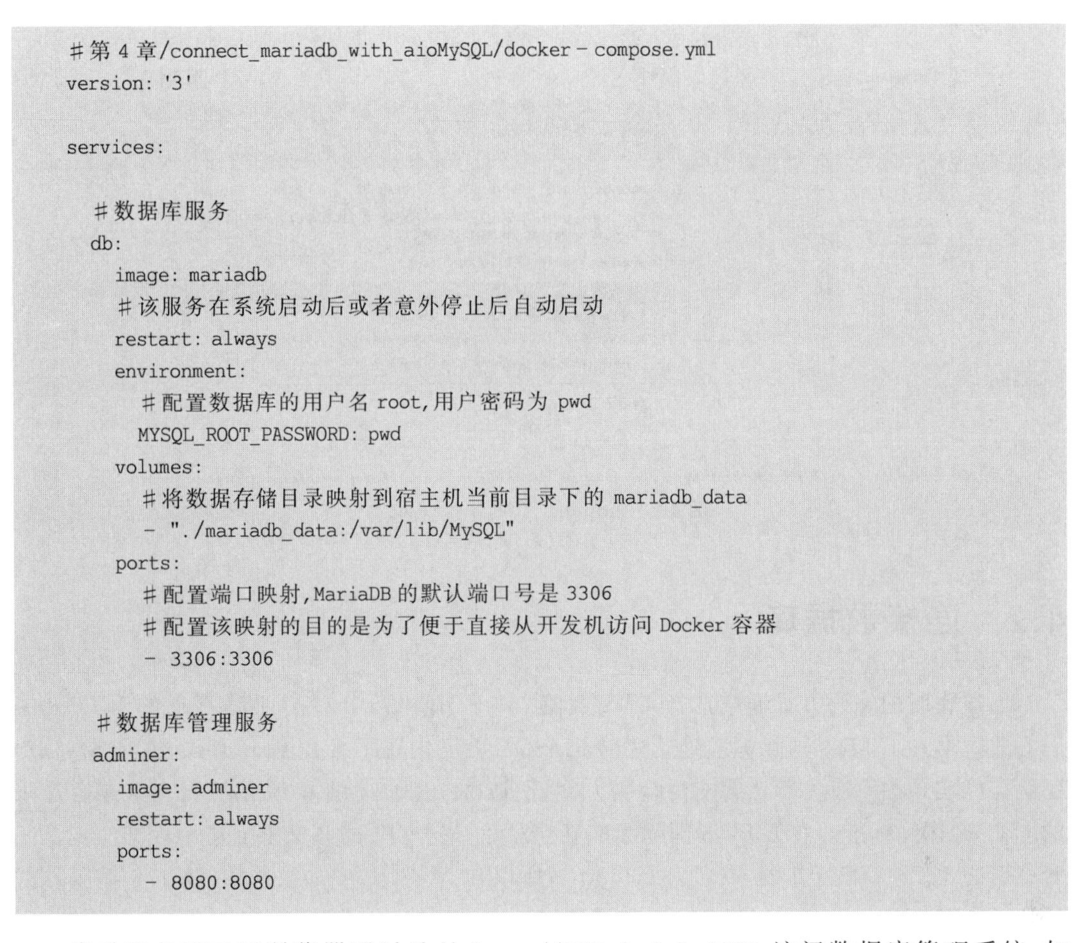

```
#第 4 章/connect_mariadb_with_aioMySQL/docker-compose.yml
version: '3'

services:

  #数据库服务
  db:
    image: mariadb
    #该服务在系统启动后或者意外停止后自动启动
    restart: always
    environment:
      #配置数据库的用户名 root,用户密码为 pwd
      MYSQL_ROOT_PASSWORD: pwd
    volumes:
      #将数据存储目录映射到宿主机当前目录下的 mariadb_data
      - "./mariadb_data:/var/lib/MySQL"
    ports:
      #配置端口映射,MariaDB 的默认端口号是 3306
      #配置该映射的目的是为了便于直接从开发机访问 Docker 容器
      - 3306:3306

  #数据库管理服务
  adminer:
    image: adminer
    restart: always
    ports:
      - 8080:8080
```

启动服务后可用浏览器通过地址 http://127.0.0.1:8080 访问数据库管理系统，如图 4-1 所示。

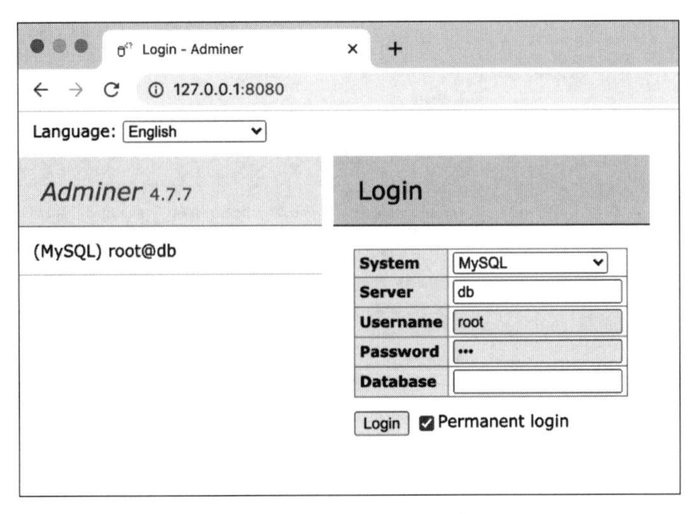

图 4-1　Adminer 数据库管理系统登录界面

输入用户名 root 和密码 pwd 以登录该管理系统，如图 4-2 所示。

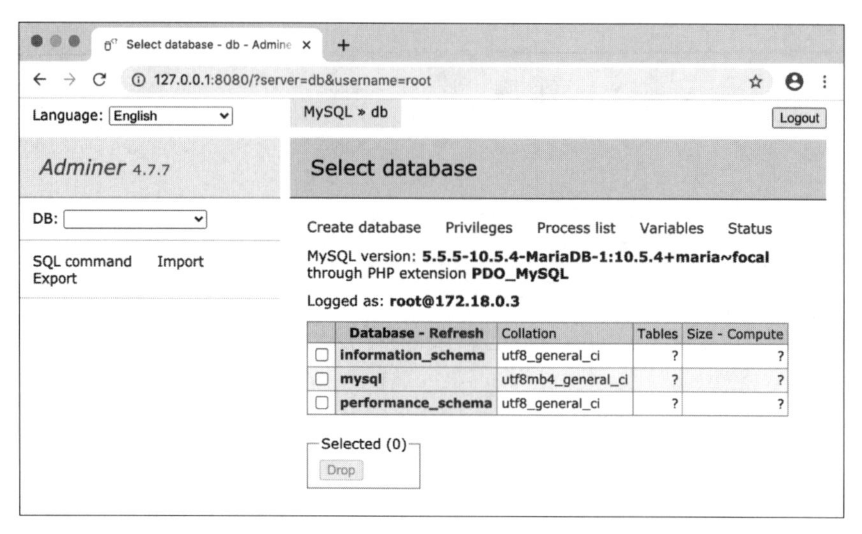

图 4-2　Adminer 数据库管理系统

4.2　连接数据库

创建库时的编码方式通常选择 UTF-8，但 MySQL 中的 utf8 并不是真正的 UTF-8 编码方式。早在 UTF-8 国际标准确立之前 MySQL 就已内置了 utf8 编码方式，这是 MySQL 团队自己实现的一套国际语言支持解决方案，所以 MySQL 中的 utf8 并不能完全兼容后来的 UTF-8 国际标准。在 UTF-8 国际标准确立之后，MySQL 团队又根据这个标准实现了一套全新的 UTF-8 兼容方案，取名为 utf8mb4，所以在 MySQL 中如果采用 UTF-8 编码方式须选择 utf8mb4。

使用 Adminer 创建一个库名为 mydb，并采用 utf8mb4 编码方式，如图 4-3 和图 4-4 所示。

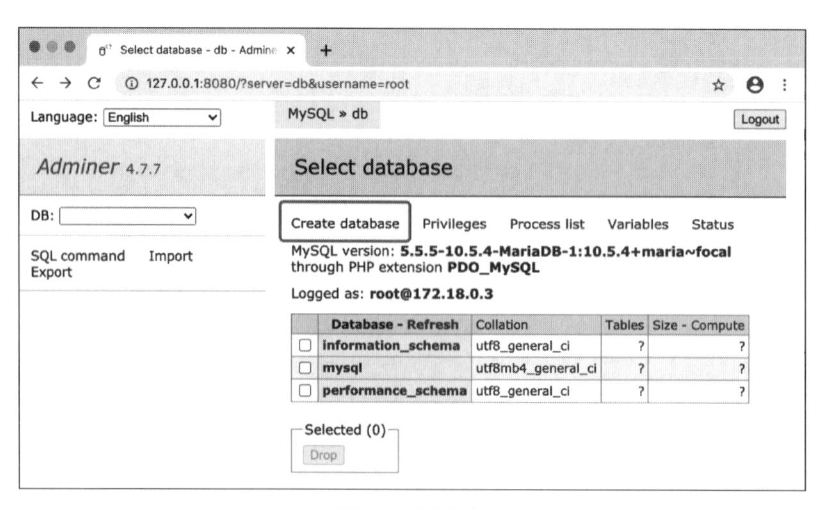

图 4-3　创建库

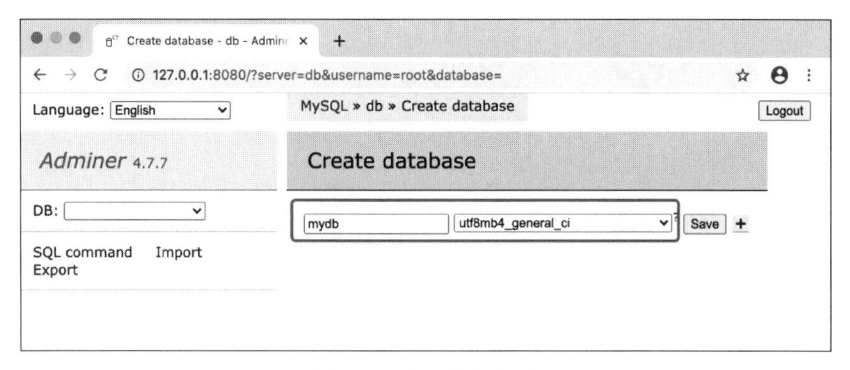

图 4-4 选择编码方式

接下来创建一个名为 student 的表，其中包括自增 id 列、名字列、年龄列，如图 4-5～图 4-7 所示。在创建表的过程中，务必将所有配置与图例保持一模一样。

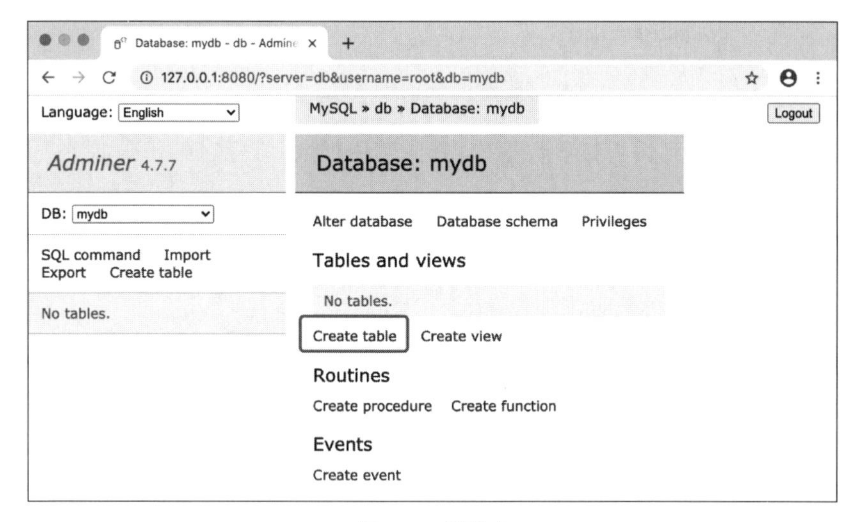

MySQL » db » mydb » Create table　　Logout

Create table

Table name: student　(engine)　(collation)

Save

Column name	Type	Length	Options	NULL	AI
id	int		unsigned	☐	◉
student_name	varchar	512	(collation)	☑	○
student_age	int			☑	○
	int			☑	○

Auto Increment:　☐Default values ☐Comment

Save

Partition by

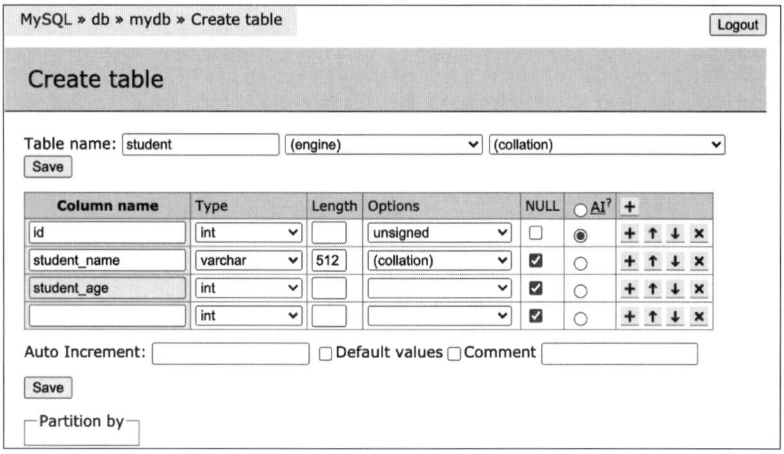

图 4-6 字段及属性选择

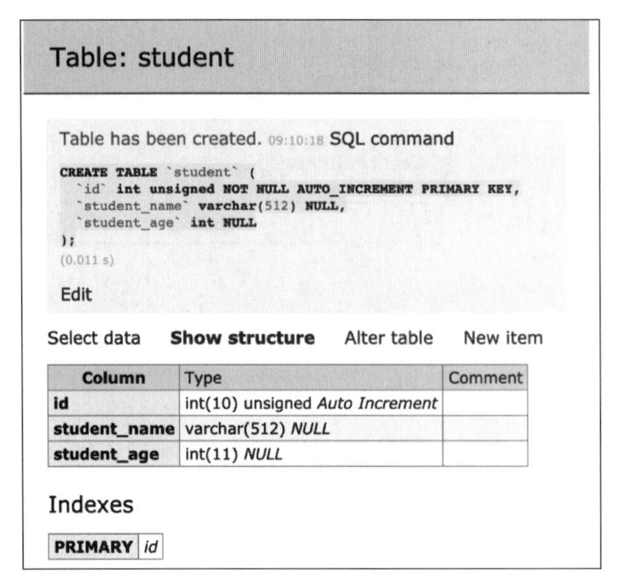

图 4-7　表创建完成

下面使用 aioMySQL 去连接该数据库，并向其中插入一条数据，代码如下：

```
"""
第 4 章/connect_mariadb_with_aioMySQL/app.py
"""
import asyncio, aioMySQL

async def main():
    #建立与数据库的连接
    conn: aioMySQL.Connection = await aioMySQL.connect(
        host = '127.0.0.1', port = 3306, user = 'root',
        password = 'pwd', db = 'mydb'
    )
    #创建一个 cursor 对象，用于操作数据库
    cur: aioMySQL.Cursor = await conn.cursor()
    #执行一条 SQL 语句，返回值是受影响的数据的条数
    effected = await cur.execute(
        "INSERT INTO `student` (`student_name`, `student_age`) "
        "VALUES ('小云', 10);"
    )
    print(effected)
    #将更改提交到数据库
    await conn.commit()
    #关闭 cursor 对象
    await cur.close()
    #关闭连接对象
    conn.close()


if __name__ == '__main__':
    asyncio.run(main())
```

代码运行的逻辑顺序是先建立连接，再基于该连接创建一个 cursor 对象用于执行 SQL 语句。这段代码运行后将在 student 表中插入一条数据，如图 4-8 所示。

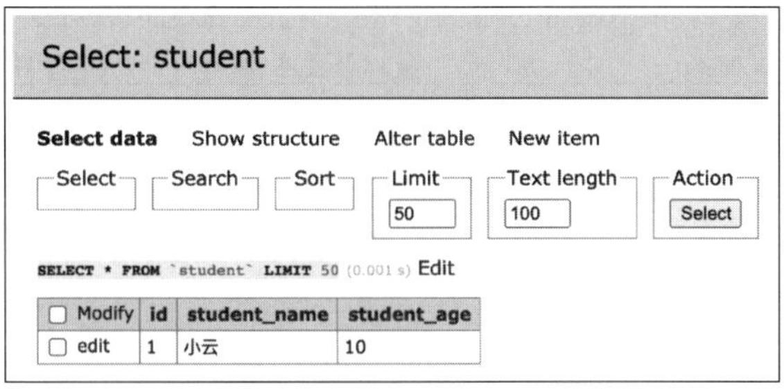

图 4-8 数据已保存到表里

为了控制与数据库的并发连接数量，还可以使用连接池实现，代码如下：

```python
"""
第 4 章/connect_mariadb_with_aioMySQL_pool/app.py
"""
import asyncio, aioMySQL


async def main():
    # 创建一个连接池，允许的最大连接数是 10，最小连接数是 0
    pool: aioMySQL.Pool = await aioMySQL.create_pool(
        minsize=0, maxsize=10,
        host='127.0.0.1', port=3306, user='root',
        password='pwd', db='mydb'
    )

    # 通过连接池启用一个连接，如果当前连接池已满，则等待其他
    # 连接被释放后再建立连接
    conn: aioMySQL.Connection = await pool.acquire()
    # 创建一个 cursor 对象用于操作数据库
    cur: aioMySQL.Cursor = await conn.cursor()
    # 执行一条 SQL 语句，返回值是受影响的数据的条数
    effected = await cur.execute(
        "INSERT INTO `student` (`student_name`, `student_age`) "
        "VALUES ('小云', 10);"
    )
    print(effected)

    await conn.commit() # 将更改提交到数据库
    await cur.close() # 关闭 cursor 对象
    pool.release(conn) # 在使用完毕后释放该连接
    pool.close()
    await pool.wait_closed()


if __name__ == '__main__':
    asyncio.run(main())
```

在一个系统中，只需创建一个连接池，用该连接池去管理所有连接，这样高效而易用。

4.3 操作数据库

11min

数据库有四大基本操作——增、删、改、查(CRUD),这一节将介绍这四大基本操作。查询出满足条件的数据,示例代码如下:

```
"""
第 4 章/aioMySQL_crud/simple_query.py
"""

import asyncio, aioMySQL


async def main():
    #创建连接池
    pool: aioMySQL.Pool = await aioMySQL.create_pool(
        minsize = 0, maxsize = 10,
        host = '127.0.0.1', port = 3306, user = 'root',
        password = 'pwd', db = 'mydb'
    )
    #启用一个连接
    conn: aioMySQL.Connection = await pool.acquire()
    #创建一个 cursor 对象,用于操作数据库
    cur: aioMySQL.Cursor = await conn.cursor()

    await cur.execute(
        "SELECT * FROM `student` WHERE `id` = '1';"
    )
    print(await cur.fetchall())
    #关闭 cursor 对象
    await cur.close()
    #释放一个连接
    pool.release(conn)
    pool.close()
    await pool.wait_closed()


if __name__ == '__main__':
    asyncio.run(main())
```

运行结果如图 4-9 所示。

```
app ×
/Users/yunp/Work/gitroot/github/plter/PythonAIOProgramming/aiomysql_crud/venv/bin/
((1, '小云', 20), (2, '小云云', 21), (3, '云加', 22), (4, '陈云', 23), (5, '杨阳', 24)

Process finished with exit code 0
```

图 4-9　查询结果

从结果中可以看出，数据直接以二维元组的方式呈现出来，这样在实际开发工作中使用起来会比较困难，所有操作需要将数据进行结构化，那么首先来分析一下表结构，通过如下代码可打印出表的字段信息：

```
await cur.execute(
    "SELECT * FROM `student` WHERE `id` = '1';"
)
print(cur.description)
```

打印出的信息如图 4-10 所示。

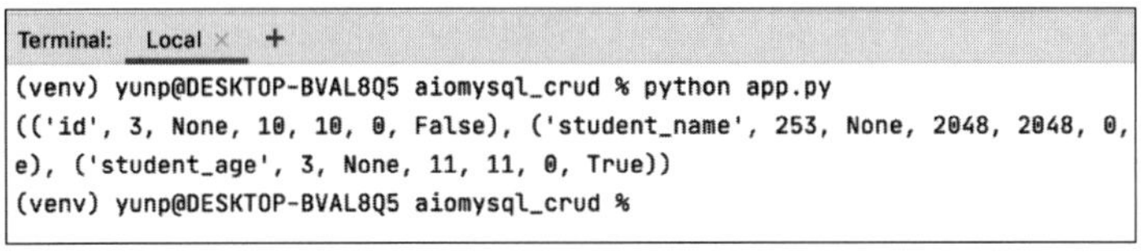

```
Terminal:  Local ×  +
(venv) yunp@DESKTOP-BVAL8Q5 aiomysql_crud % python app.py
(('id', 3, None, 10, 10, 0, False), ('student_name', 253, None, 2048, 2048, 0,
e), ('student_age', 3, None, 11, 11, 0, True))
(venv) yunp@DESKTOP-BVAL8Q5 aiomysql_crud %
```

图 4-10 表字段信息

可以看出，每个字段信息也是一个元组，元组中的第一项是字段的名字，这就是我们所需要的，将数据结果与字段名映射即可形成结构化的数据，通过如下代码可将结果数据转换成包含多个对象的数组：

```
#获取原始数据结果
raw_data = await cur.fetchall()
#计算字段个数
field_range = range(len(cur.description))
#将字段名与原始结果映射生成对象数组
result = [
    {cur.description[i][0]: row[i] for i in field_range}
    for row in raw_data
]
print(result)
```

打印输出的信息如图 4-11 所示。

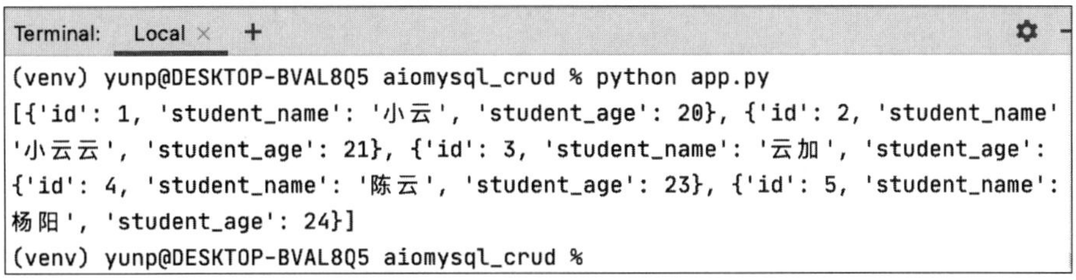

```
Terminal:  Local ×  +
(venv) yunp@DESKTOP-BVAL8Q5 aiomysql_crud % python app.py
[{'id': 1, 'student_name': '小云', 'student_age': 20}, {'id': 2, 'student_name'
'小云云', 'student_age': 21}, {'id': 3, 'student_name': '云加', 'student_age':
{'id': 4, 'student_name': '陈云', 'student_age': 23}, {'id': 5, 'student_name':
杨阳', 'student_age': 24}]
(venv) yunp@DESKTOP-BVAL8Q5 aiomysql_crud %
```

图 4-11 将结果结构化

修改一条数据的代码如下：

```
await print_all_data(cur)
await cur.execute(
```

```
        "UPDATE `student` SET "
        "`student_name` = '小云加', "
        "`student_age` = '22' "
        "WHERE `id` = '1';"
    )
    await print_all_data(cur)
```

则两次输出的结果所显示的数据修改对比，如图 4-12 所示。

```
[{'id': 1, 'student_name': '小云', 'student_age': 20},
[{'id': 1, 'student_name': '小云加', 'student_age': 22}
```

图 4-12 修改数据

删除一条数据代码如下：

```
await cur.execute(
    "DELETE FROM `student` WHERE `id` = '1';"
)
```

完整的示例代码如下：

```
"""
第 4 章/aioMySQL_crud/app.py
"""

import asyncio, aioMySQL


async def print_all_data(cur):
    await cur.execute(
        "SELECT * FROM `student` WHERE id > 0;"
    )
    #获取原始数据结果
    raw_data = await cur.fetchall()
    #计算字段个数
    field_range = range(len(cur.description))
    #将字段名与原始结果映射生成对象数组
    result = [
        {cur.description[i][0]: row[i] for i in field_range}
        for row in raw_data
    ]
    print(result)


async def main():
    #创建连接池
    pool: aioMySQL.Pool = await aioMySQL.create_pool(
        minsize = 0, maxsize = 10,
        host = '127.0.0.1', port = 3306, user = 'root', password = 'pwd', db = 'mydb'
```

```
    )
    conn: aioMySQL.Connection = await pool.acquire() #启用一个连接
    cur: aioMySQL.Cursor = await conn.cursor() #创建一个 cursor 对象,用于操作数据库

    await print_all_data(cur)
    await cur.execute(
        "UPDATE`student` SET "
        "`student_name` = '小云加', "
        "`student_age` = '22' "
        "WHERE`id` = '1';"
    )
    await print_all_data(cur)

    #await print_all_data(cur)
    #await cur.execute(
    # "DELETE FROM `student` WHERE`id` = '1';"
    #)
    #await print_all_data(cur)

    await conn.commit()  #将更改提交到数据库
    await cur.close()  #关闭 cursor 对象
    pool.release(conn)  #释放一个连接
    pool.close()
    await pool.wait_closed()


if __name__ == '__main__':
    asyncio.run(main())
```

4.4 SQLAlchemy 异步

虽然 aioMySQL 可以直接操作数据库,而且还很方便,但是总有一部分人认为在编程语言中写 SQL 语句容易造成混乱,最重要的是,代码不够优雅。你可能会笑,难道代码不够优雅也是问题吗?是的,而且是很大的问题,如果项目很庞大,不优雅的代码意味着可读性差,团队协作的成本会提高,而且不利于代码的维护与更新。一个优秀的项目,可能不是运行速度最快的,但一定是编码风格优秀的。

所以出现了一系列优秀的框架,使不在程序语言中出现 SQL 也能操作数据库,SQLAlchemy 就是其中之一。

aioMySQL 直接内置了 SQLAlchemy 所支持的功能,在使用之前,需要通过代码声明数据库的结构,代码如下:

```
student = sqlalchemy.Table(
    'student', sqlalchemy.MetaData(),
    sqlalchemy.Column('id', sqlalchemy.Integer, primary_key=True),
```

```
    sqlalchemy.Column('student_name', sqlalchemy.String(255)),
    sqlalchemy.Column('student_age', sqlalchemy.Integer)
)
```

之后在程序中才能使用该表，并可查询出表中的所有数据，示例代码如下：

```
result = await conn.execute(student.select())
all_data = await result.fetchall()
print(all_data)
```

使用 aioMySQL 内置的 SQLAlchemy 模块进行增、删、改、查的操作的完整示例代码如下：

```
"""
第 4 章/aioMySQL_sqlalchemy/app.py
"""
import asyncio, aioMySQL.sa, sqlalchemy, aioMySQL.sa.result

student = sqlalchemy.Table(
    'student', sqlalchemy.MetaData(),
    sqlalchemy.Column('id', sqlalchemy.Integer, primary_key = True),
    sqlalchemy.Column('student_name', sqlalchemy.String(255)),
    sqlalchemy.Column('student_age', sqlalchemy.Integer)
)

async def print_all_data(conn):
    result: aioMySQL.sa.result.ResultProxy = await conn.execute(
        student.select()
    )
    all_data = await result.fetchall()
    print(all_data)


async def main():
    engine: aioMySQL.sa.Engine = await aioMySQL.sa.create_engine(
        host = '127.0.0.1', port = 3306, user = 'root',
        password = 'pwd', db = 'mydb'
    )
    conn: aioMySQL.sa.SAConnection = await engine.acquire()

    #增加一条数据
    await conn.execute(student.insert().values(
        student_name = '杨阳', student_age = 20
    ))
    await print_all_data(conn)

    #修改一条数据
    await conn.execute(student.update().where(
        student.columns.id == 1
```

```
        ).values(student_age = 10))
        await print_all_data(conn)

        # 删除一条数据
        await conn.execute(student.delete().where(student.columns.id == 1))
        await print_all_data(conn)
        engine.release(conn)


if __name__ == '__main__':
    asyncio.run(main())
```

运行结果如图 4-13 所示。

```
app ×
/Users/yunp/Work/gitroot/github
[(1, '杨阳', 20)]
[(1, '杨阳', 10)]
[]

Process finished with exit code
```

图 4-13 数据操作运行结果

4.5 与 AIOHTTP 集成

在这一节里,以一个综合实例来完整介绍如何在网站开发中集成数据库。项目最终主页效果如图 4-14 所示。

这是一个简单的学生管理系统,实现了基本的增、删、改、查功能,单击右上角"+"按钮可以进入添加学生数据的界面,如图 4-15 所示。

学生列表 +

id	姓名	年龄	
10	杨明	10	编辑 删除
11	小云	20	编辑 删除
12	云加	21	编辑 删除
13	小明	12	编辑 删除

图 4-14 数据表格

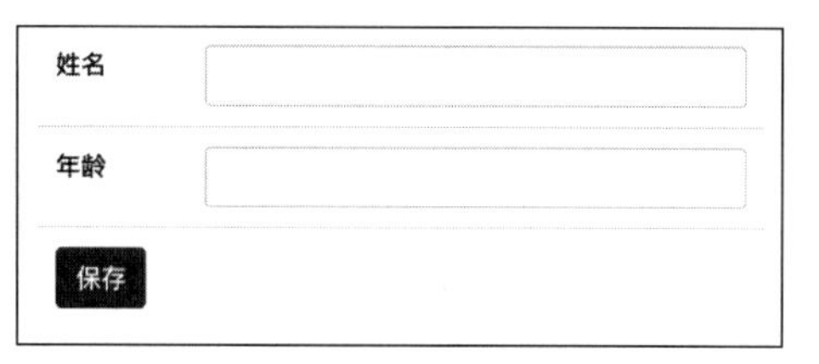

图 4-15 添加数据

单击"编辑"按钮可以对某一条数据进行编辑,如图 4-16 所示。

单击"删除"按钮将弹出确认对话框,在用户确认后可以删除对应的数据。如图 4-17 所示。

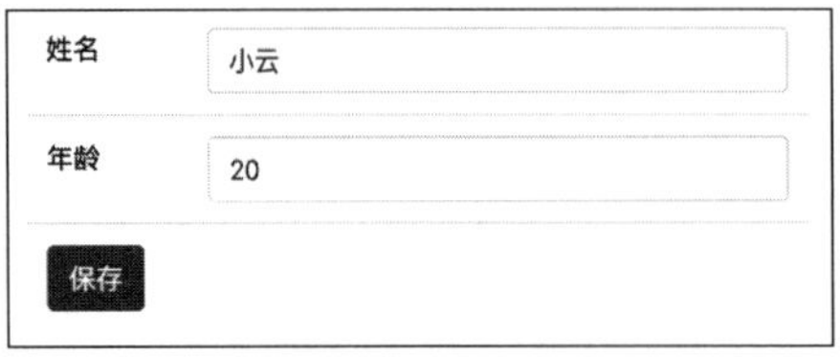

图 4-16　编辑数据

图 4-17　删除数据前提示

关于前端框架的配置，我们在 3.4 节模板渲染中已经讲过，本节不再赘述，为了方便阅读，笔者先把项目的文件结构截图以供参考，如图 4-18 所示。

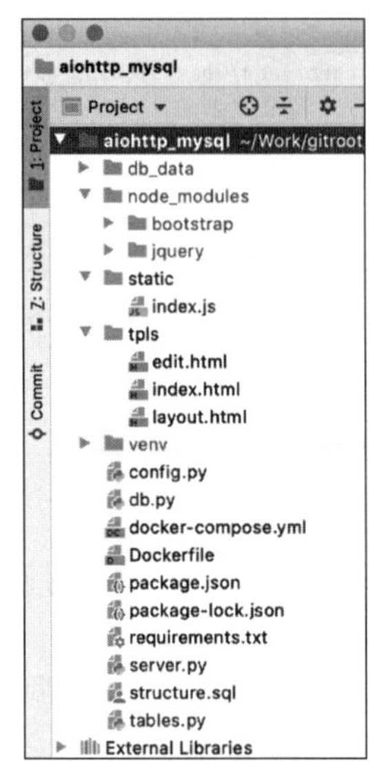

图 4-18　项目文件结构

在实现该项目的过程中，一定要做到模块化。首先把项目中可能用到的所有配置信息抽象出来，然后创建一个单独的文件并命名为 config.py，其中代码如下：

```
"""
第 4 章/aiohttp_MySQL/config.py
"""
import os

#服务器根目录
SERVER_ROOT = os.path.dirname(__file__)

#所有静态文件目录
STATIC_MAPPING = [
    dict(
        web_path = "/node_modules",
        dir = os.path.join(SERVER_ROOT, "node_modules")
```

```
    ),
    dict(
        web_path = "/static",
        dir = os.path.join(SERVER_ROOT, "static")
    )
]

#模板文件根目录
TEMPLATE_ROOT = os.path.join(SERVER_ROOT, "tpls")

#数据库连接相关配置信息
DB_HOST = 'db'
DB_PORT = 3306
DB_NAME = 'mydb'
DB_USER = 'root'
DB_PASSWORD = 'pwd'
```

接下来把数据库连接这部分功能也写成一个单独的文件 db.py，并且为请求处理函数提供一个装饰器函数用于更方便地支持数据库操作，代码如下：

```
"""
第 4 章/aiohttp_MySQL/db.py
"""
from aioMySQL.sa import create_engine, Engine
import config

file_scope_vars = {}


async def get_engine():
    """
获取数据库引擎单例
    :return:
    """
    if "engine" not in file_scope_vars:
        file_scope_vars['engine'] = await create_engine(
            host = config.DB_HOST,
            port = config.DB_PORT,
            user = config.DB_USER,
            password = config.DB_PASSWORD,
            db = config.DB_NAME
        )
    return file_scope_vars['engine']


def with_db(fun):
    """
用于给请求处理函数添加数据库支持的装饰器函数
    :param fun:
```

```
    :return:
    """

    async def wrapper(req):
        # 获取数据库引擎
        engine: Engine = await get_engine()
        # 创建一个数据库连接实例
        db = await engine.acquire()
        try:
            result = await fun(req, db)
            # 执行结束后释放数据库连接
            engine.release(db)
            return result
        except Exception as e:
            # 捕获异常后释放数据库连接
            engine.release(db)
            raise e

    return wrapper
```

get_engine 函数的内部实现了单例模式，目的是为了不管从哪里调用该函数，保证返回的结果是同一个 aioMySQL. sa. engine. Engine 实例。

with_db 函数是一个用于请求处理函数的装饰器，目的是为了给请求处理函数提供一个 db 参数，用于操作数据库，而在请求处理函数内部无须关心 db 的创建与释放。

接下来创建一个 tables. py 文件，并把所有与数据库相关的声明都写在这个文件中，当然 student 表结构也在这个文件中，代码如下：

```
import sqlalchemy

# 定义表结构
student = sqlalchemy.Table(
    'student', sqlalchemy.MetaData(),
    sqlalchemy.Column('id', sqlalchemy.Integer, primary_key = True),
    sqlalchemy.Column('student_name', sqlalchemy.String(255)),
    sqlalchemy.Column('student_age', sqlalchemy.Integer)
)
```

为了便于理解，这里同时提供了对应的数据库备份文件，代码如下：

```
-- 第 4 章/aiohttp_MySQL/structure.sql

SET NAMES utf8;
SET time_zone = '+00:00';
SET foreign_key_checks = 0;
SET sql_mode = 'NO_AUTO_VALUE_ON_ZERO';
```

```
SET NAMES utf8mb4;

CREATE DATABASE `mydb` /*!40100 DEFAULT CHARACTER SET utf8mb4 */;
USE `mydb`;

DROP TABLE IF EXISTS `student`;
CREATE TABLE `student` (
  `id` int(10) unsigned NOT NULL AUTO_INCREMENT,
  `student_name` varchar(512) DEFAULT NULL,
  `student_age` int(11) DEFAULT NULL,
  PRIMARY KEY (`id`)
) ENGINE = InnoDB DEFAULT CHARSET = utf8mb4;
```

接下来实现程序主文件 server.py，代码如下：

```
"""
第 4 章/aiohttp_MySQL/server.py
"""
from aiohttp import web
from db import with_db
from aioMySQL.sa import SAConnection, result
import aiohttp_jinja2, jinja2, config, tables

routes = web.RouteTableDef()


@routes.get('/')
@aiohttp_jinja2.template("index.html")
@with_db
async def index(req, db: SAConnection):
    # 实现的功能是查询出所有学生并呈现出来
    exec_result: result.ResultProxy = await db.execute(
        tables.student.select()
    )
    data = await exec_result.fetchall()
    return dict(students = data, title = "学生列表")


@routes.get('/edit')
@aiohttp_jinja2.template("edit.html")
@with_db
async def edit(req: web.Request, db: SAConnection):
    """
编辑页面,我们把编辑功能和添加功能放在一起实现,如果页面没
有传入 id 参数,则把该页面当成添加学生信息页面对待,如果传入
了 id 参数,则当成编辑学生信息页面对待
    """
    student_id = req.query.getone("id") if "id" in req.query else None
    student = None
```

```
    if student_id: # 如果页面传入 student_id,则启用编辑,否则执行添加操作
        student_result: result.ResultProxy = await db.execute(
            tables.student.select().where(
                tables.student.columns.id == student_id
            )
        )
        student = await student_result.fetchone()
    return dict(title = "编辑", student = student)


@routes.post('/edit')
@with_db
async def edit(req: web.Request, db: SAConnection):
    """
处理表单提交的页面,如果没有传入 id,则执行添加学生信息的操作,如果
传入了 id,则根据 id 判断指定的学生是否存在,如果存在,更新该学生
的信息,如果不存在则添加该学生.在处理完成后跳转到首页
    """
    params = await req.post()
    student_name = params['student_name'] \
        if "student_name" in params else None
    student_age = params['student_age'] \
        if "student_age" in params else None
    student_id = params['student_id'] \
        if "student_id" in params else None
    if not student_name or not student_age:
        return web.Response(text = "Parameters error")
    if student_id: # 如果有 student_id,则尝试查找这条数据
        ret: result.ResultProxy = await db.execute(
            tables.student.select().where(
                tables.student.columns.id == student_id
            )
        )
        if ret.rowcount: # 如果存在这条记录,则更新这条记录
            conn = await db.begin()
            await db.execute(
                tables.student.update()
                    .where(tables.student.columns.id == student_id)
                    .values(student_name = student_name,
                            student_age = student_age)
            )
            await conn.commit()
            raise web.HTTPFound("/")
    # 能够执行到这里,说明指定 student_id 的记录不存在或者没
    # 有指定 student_id,此时执行添加新数据操作
    conn = await db.begin()
    await db.execute(
        tables.student.insert()
            .values(student_name = student_name, student_age = student_age)
    )
```

```
    await conn.commit()
    raise web.HTTPFound("/")


@routes.get('/remove')
@with_db
async def remove(req: web.Request, db: SAConnection):
    """
处理删除学生数据的页面,根据传入的 id 删除指定的学生数据
删除后跳转到首页
    """
    student_id = req.query.getone("id") if "id" in req.query else None
    if student_id:
        conn = await db.begin()
        await db.execute( # 根据 student_id 删除数据
            tables.student
                .delete()
                .where(tables.student.columns.id == student_id)
        )
        await conn.commit()
        raise web.HTTPFound("/")
    else:
        return web.Response(text = "Parameters error")


if __name__ == '__main__':
    app = web.Application()
    # 配置模板文件根目录
    aiohttp_jinja2.setup(
        app,
        loader = jinja2.FileSystemLoader(config.TEMPLATE_ROOT)
    )
    app.add_routes(routes)
    # 配置静态文件目录
    for m in config.STATIC_MAPPING:
        app.router.add_static(m['web_path'], m['dir'])
    web.run_app(app, port = 8000)
```

与首页对应的模板文件 index.html 的内容如下：

```
<!-- 第 4 章/aiohttp_MySQL/tpls/index.html -->
{ % extends "layout.html" % }
{ % block body % }
<div class = "card" style = "margin - top: 2rem;">
<div class = "card - header" style = "position: relative;">
<a href = "/edit"
        style = "position: absolute;right:1rem;top:0;font - size: 20pt">
         +
</a>
```

```
学生列表
</div>
<div>
<table class="table table-striped" style="margin-bottom: 0">
<thead>
<tr>
<th>id</th>
<th>姓名</th>
<th>年龄</th>
<th></th>
</tr>
</thead>
<tbody>
                {% for s in students %}
<tr>
<td>{{ s.id }}</td>
<td>{{ s.student_name }}</td>
<td>{{ s.student_age }}</td>
<td style="width: 10rem;text-align: center;">
<a href="/edit?id={{ s.id }}"
                                class="btn btn-primary btn-sm">
编辑
</a>
<a href="/remove?id={{ s.id }}"
                                class="btn-delete-item btn btn-danger btn-sm">
删除
</a>
</td>
</tr>
                {% endfor %}
</tbody>
</table>
</div>
</div>
<script src="/static/index.js"></script>
{% endblock %}
```

在 index.html 中引用的 index.js 文件内容如下：

```
//第 4 章/aiohttp_MySQL/static/index.js
(function () {
    $(".btn-delete-item").click(function (e) {
        if (!confirm("你真的要删除这条数据吗?")) {
            e.preventDefault();
        }
    });
})();
```

与编辑页面对应的模板文件 edit.html 的内容如下：

```
<!-- 第 4 章/aiohttp_MySQL/tpls/edit.html -->
{% extends "layout.html" %}
{% block body %}
<form method = "post" enctype = "application/x-www-form-URLencoded">
        {% if student %}
<input type = "hidden" value = "{{ student.id }}"
                    name = "student_id">
        {% endif %}
<table class = "table">
<tbody>
<tr>
<td>姓名</td>
<td>
<input type = "text" required class = "form-control"
                              name = "student_name"
                    value = "{{ student.student_name if student else '' }}">
</td>
</tr>
<tr>
<td>年龄</td>
<td>
<input type = "number" required class = "form-control"
                              name = "student_age"
                    value = "{{ student.student_age if student else '' }}">
</td>
</tr>
</tbody>
<tfoot>
<tr>
<td colspan = "2">
<input class = "btn btn-primary" type = "submit"
                              value = "保存">
</td>
</tr>
</tfoot>
</table>
</form>
{% endblock %}
```

根模板文件 layout.html 的内容如下：

```
<!-- 第 4 章/aiohttp_MySQL/tpls/layout.html -->
<!DOCTYPE html>
<html lang = "en">
<head>
<meta charset = "UTF-8">
<title>{{ title }}</title>
```

```
<link rel="stylesheet" href="/node_modules/bootstrap/dist/css/bootstrap.min.css">

<script src="/node_modules/jquery/dist/jquery.min.js"></script>
<script src="/node_modules/bootstrap/dist/js/bootstrap.bundle.min.js"></script>
</head>
<body>
<div class="container">
    {% block body %}
    {% endblock %}
</div>
</body>
</html>
```

构建该项目镜像的 Dockerfile 文件内容如下：

```
#第4章/aiohttp_MySQL/Dockerfile
FROM python:3.8-slim
RUN pip install -i https://pypi.tuna.tsinghua.edu.cn/simple aiohttp
RUN pip install -i https://pypi.tuna.tsinghua.edu.cn/simple aioMySQL
RUN pip install -i https://pypi.tuna.tsinghua.edu.cn/simple sqlalchemy
RUN pip install -i https://pypi.tuna.tsinghua.edu.cn/simple aiohttp_jinja2
```

Docker 服务配置文件 docker-compose.yml 的内容如下：

```
#第4章/aiohttp_MySQL/docker-compose.yml
version: "3"

services:
  web:
    build: .
    ports:
      - "80:8000"
    restart: always
    volumes:
      - ".:/opt/"
    working_dir: "/opt/"
    command: python3 ./server.py

  db:
    image: mariadb
    restart: always
    environment:
      MYSQL_ROOT_PASSWORD: pwd
    volumes:
      - "./db_data:/var/lib/MySQL"
    ports:
      - 3306:3306

  adminer:
```

```
    image: adminer
    restart: always
    ports:
      - 8080:8080
```

从服务配置文件中可以看出，在集成开发环境中开发调试时使用 8000 端口，正式部署到 Docker 环境中使用的是 80 端口。

第5章 ASGI

ASGI(Asynchronous Server Gateway Interface)为异步服务网关接口，秉承 WSGI 统一网关接口原则，在异步服务、框架和应用之间提供一个标准接口，同时兼容 WSGI。

学习一个新知识之前，我们首先需要明白它为什么会出现，那么为什么 ASGI 和 WSGI 这类的统一网关接口会出现呢？

Python Web 开发大致可以分为两大层，协议层(HTTP 逻辑层，后简称协议层)和应用层，协议层主要根据 HTTP 协议解码与编码数据，应用层主要实现具体的业务逻辑，分层的好处是应用层可以根据需要很方便地替换协议层，要做到这一点，只需为协议层和应用层设计一个统一的接口，这就是 WSGI 与 ASGI。

有了这个统一的接口，协议层不仅仅只有 Python 才能实现，其他的语言例如 C 语言如果按这个接口去实现协议层，那么也可以和 Python 应用层对接，所以出现的 WSGI 可让 Python 运行在 Apache 服务器上，这样可以让 Apache 强大的静态文件请求处理功能发挥作用，避免 Python 去处理静态请求，从而大大提高整个网站的运行速度。

6min

5.1 WSGI

WSGI 可以与传统阻塞型 IO 服务器配合使用。

在这里需要明确一点，传统阻塞型 IO 服务器就一定是低效的吗？当然不是，我们已知异步 IO 编程模型的目的是为了合理利用 IO 资源，而传统的阻塞型 IO 编程也可以通过复杂的方案实现合理利用 IO 资源，例如优秀的 Apache 服务器。当然最新的 Apache 服务器也采用了异步 IO 编程模型，毕竟是操作系统已经实现了的功能，直接使用能够让项目的维护成本降低。

在传统阻塞型 IO 编程模型中，通常会把 Python 应用服务与 Apache 服务器结合使用，仅把与业务逻辑处理相关的任务交给 Python 语言去完成，而其他资源的处理(例如：静态文件请求处理、代理等)任务交给强大的 Apache 服务器去完成，从而实现负载均衡，提高项目整体的运行效率。

接下来我们以搭建一个 Apache + WSGI 的 Docker 环境为例演示如何在项目中使用 WSGI。项目结构如图 5-1 所示。

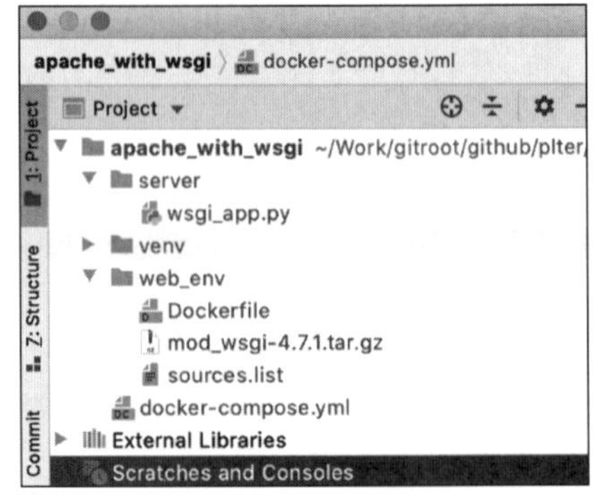

图 5-1 项目结构

先从编写 Docker 构建脚本开始介绍，我们计划使用的基础镜像是 httpd:2.4。因为需要重新编译 mod_wsgi 模块，所以需要下载庞大的 C/C++编译环境，而官方镜像服务器在国外，下载速度较慢，所以这里使用清华的 Debian 镜像。

首先需要准备一个 sources.list 文件，其内容如下：

```
#第 5 章/apache_with_wsgi/web_env/sources.list

deb http://mirrors.tuna.tsinghua.edu.cn/debian/ buster main contrib non-free
#deb-src https://mirrors.tuna.tsinghua.edu.cn/debian/ buster main contrib
#non-free
deb http://mirrors.tuna.tsinghua.edu.cn/debian/ buster-updates main contrib non-free
#deb-src https://mirrors.tuna.tsinghua.edu.cn/debian/ buster-updates main
#contrib non-free
deb http://mirrors.tuna.tsinghua.edu.cn/debian/ buster-backports main
contrib non-free
#deb-src https://mirrors.tuna.tsinghua.edu.cn/debian/ buster-backports
#main contrib non-free
deb http://mirrors.tuna.tsinghua.edu.cn/debian-security buster/updates
main contrib non-free
#deb-src https://mirrors.tuna.tsinghua.edu.cn/debian-security
#buster/updates main contrib non-free
```

mod_wsgi 的源码可从 https://GitHub.com/GrahamDumpleton/mod_wsgi 下载，在这里使用的 mod_wsgi-4.7.1，如果使用其他版本，需要注意修改相应的目录。Dockerfile 文件的内容如下：

```
#第 5 章/apache_with_wsgi/web_env/Dockerfile
FROM httpd:2.4.41

RUN mv /etc/apt/sources.list /etc/apt/sources.list.bak
#替换软件源为清华镜像站
COPY sources.list /etc/apt/
#更新软件源
RUN apt-get update
#安装 Python 运行环境及 C 语言编译器用以编译 WSGI
RUN apt-get install -y python3 python3-dev gcc g++ make
#将 WSGI 源码复制到镜像中
COPY mod_wsgi-4.7.1.tar.gz /opt
#切换镜像中的工作目录
WORKDIR /opt
#解压 wsgi 源码包
RUN tar -xzvf ./mod_wsgi-4.7.1.tar.gz
#切换工作目录到 WSGI 源码目录
WORKDIR /opt/mod_wsgi-4.7.1
#配置编译环境
RUN ./configure --with-apxs=/usr/local/apache2/bin/apxs \
    --with-python=/usr/bin/python3
#编译并安装
```

```
RUN make&&make install

# 这里对 Apache 服务器的配置文件进行修改,添加 WSGI 相关的配置项
# 包括加载 WSGI 动态库、配置脚本文件、配置脚本文件所在目录的权限
RUN echo 'LoadModule wsgi_module modules/mod_wsgi.so' >>\
    /usr/local/apache2/conf/httpd.conf
RUN echo 'WSGIScriptAlias /app /opt/server/wsgi_app.py' >>\
    /usr/local/apache2/conf/httpd.conf
RUN echo '<Directory /opt/server/>' >>\
    /usr/local/apache2/conf/httpd.conf
RUN echo ' Require all granted' >> \
    /usr/local/apache2/conf/httpd.conf
RUN echo '</Directory>' >> /usr/local/apache2/conf/httpd.conf
```

脚本文件 wsgi_app.py 的内容如下：

```
"""
第 5 章/apache_with_wsgi/server/wsgi_app.py
"""


def application(environ, start_response):
    status = '200 OK'
    output = b'Hello World!'

    response_headers = [
        ('Content-type', 'text/plain'),
        ('Content-Length', str(len(output)))
    ]
    start_response(status, response_headers)
    return [output]
```

这是一个最简单的 WSGI 脚本，其中声明了一个 application 函数，当有前端请求时，该函数会被调用以处理请求，这个示例中向前端返回'Hello World!'。

对应的 docker-compose.yml 文件内容如下：

```
#第 5 章/apache_with_wsgi/docker-compose.yml
version: "3"

services:
  web:
    build: ./web_env
    tty: true
    ports:
      - "80:80"
    volumes:
      - "./server:/opt/server"
```

在终端里执行命令 docker-compose up-d 以构建并启动该服务器，待服务器启动完成后，通过浏览器访问 http://127.0.0.1/app，显示效果如图 5-2 所示。

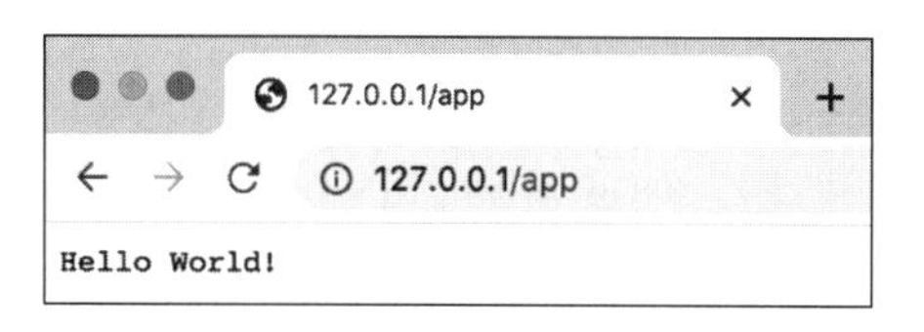

图 5-2 Apache 集成 WSGI 示例页面

用同样的方式，还可以集成第三方框架如 web.py、web2py、Django 等以提高开发效率。但这些都已经是过时的技术了，讲解 WSGI 是为了让读者拥有维护旧项目的能力，并对 ASGI 有一个初步的认知，也因为目前 Apache 和 Nginx 平台均未出现 ASGI 模块，无法演示 ASGI 模块的编译与配置，但构建 ASGI 与构建 WSGI 类似，所以只能以构建 WSGI 环境来演示构建过程。

5.2 ASGI

8min

ASGI 是根据统一接口的思想重新设计的新标准，你可能会有疑问，为什么不直接升级 WSGI 而去创造新的标准呢？

WSGI 是基于 HTTP 短连接的网关接口，一次调用请求必须尽快处理完毕并返回结果，这种模式并不适用于长连接，例如 HTML 5 新标准中的技术 SSE(Server-Sent Events)和 WebSocket，WSGI 及传统阻塞型 IO 编程模型并不擅长处理这类请求。就算强行升级 WSGI 以支持异步 IO，可是如果配套的技术(如 Apache 服务器)没有提供相应的支持也是没有意义的。既然 Python 异步 IO 编程模型已经走在了前面，那就制定一个全新的标准 ASGI 以最优雅的方式支持并使用最新的技术。

ASGI 接口是一个异步函数，它要求传入 3 个参数，分别为 scope、receive 和 send，示例代码如下：

```
async def app(scope, receive, send):
    pass
```

其中 scope 是一个字典(dict)，包括连接相关的信息，图 5-3 所示是一个请求中的 scope 所包括信息的断点调试截图。

receive 是一个异步函数，用于读取前端发来的信息，一条读取到的信息结构如下：

```
{
    'type': 'http.request',
    'body': b"",
    'more_body': False
}
```

该信息中包括 3 个字段，分别为类型(type)、内容(body)和是否还有更多内容(more_body)，其中通过 type 可以用来判断该信息是什么类型，如 HTTP 请求、生命周期、WebSocket 请求等，body 是该信息中包括的数据，此数据采用二进制格式，more_body 指明当前数据是否已经发送完毕，如果发送完毕，则 more_body 的值为 False，这样便可以用来

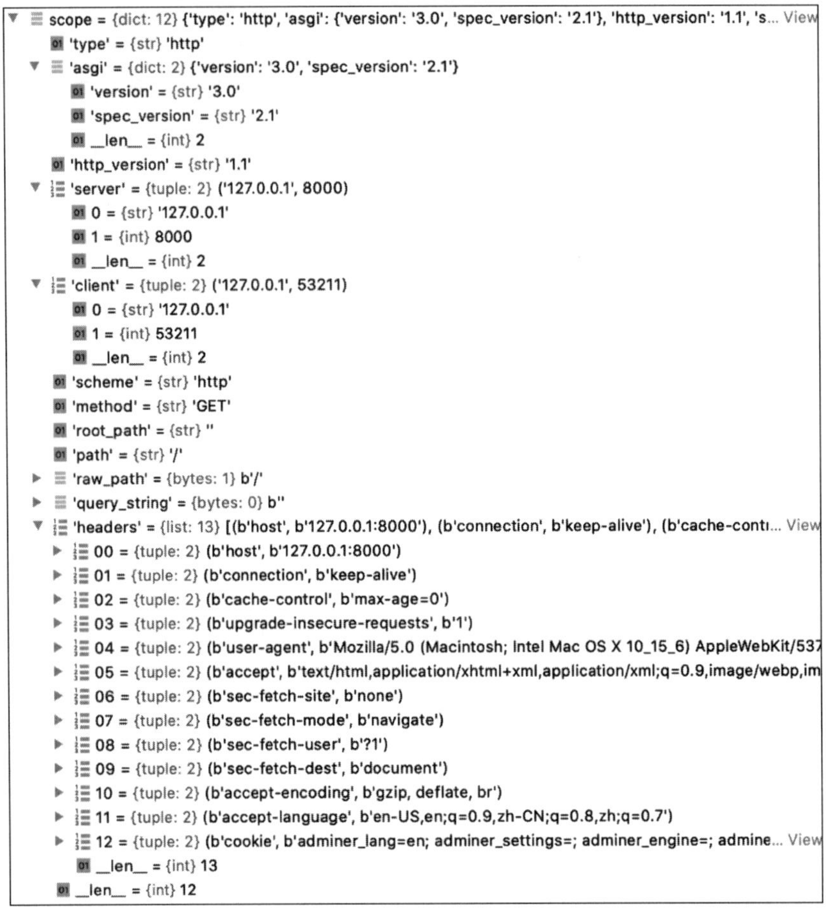

```
▼ scope = {dict: 12} {'type': 'http', 'asgi': {'version': '3.0', 'spec_version': '2.1'}, 'http_version': '1.1', 's... View
    'type' = {str} 'http'
  ▼ 'asgi' = {dict: 2} {'version': '3.0', 'spec_version': '2.1'}
      'version' = {str} '3.0'
      'spec_version' = {str} '2.1'
      __len__ = {int} 2
    'http_version' = {str} '1.1'
  ▼ 'server' = {tuple: 2} ('127.0.0.1', 8000)
      0 = {str} '127.0.0.1'
      1 = {int} 8000
      __len__ = {int} 2
  ▼ 'client' = {tuple: 2} ('127.0.0.1', 53211)
      0 = {str} '127.0.0.1'
      1 = {int} 53211
      __len__ = {int} 2
    'scheme' = {str} 'http'
    'method' = {str} 'GET'
    'root_path' = {str} ''
    'path' = {str} '/'
  ▶ 'raw_path' = {bytes: 1} b'/'
  ▶ 'query_string' = {bytes: 0} b''
  ▼ 'headers' = {list: 13} [(b'host', b'127.0.0.1:8000'), (b'connection', b'keep-alive'), (b'cache-contı... View
    ▶ 00 = {tuple: 2} (b'host', b'127.0.0.1:8000')
    ▶ 01 = {tuple: 2} (b'connection', b'keep-alive')
    ▶ 02 = {tuple: 2} (b'cache-control', b'max-age=0')
    ▶ 03 = {tuple: 2} (b'upgrade-insecure-requests', b'1')
    ▶ 04 = {tuple: 2} (b'user-agent', b'Mozilla/5.0 (Macintosh; Intel Mac OS X 10_15_6) AppleWebKit/537
    ▶ 05 = {tuple: 2} (b'accept', b'text/html,application/xhtml+xml,application/xml;q=0.9,image/webp,im
    ▶ 06 = {tuple: 2} (b'sec-fetch-site', b'none')
    ▶ 07 = {tuple: 2} (b'sec-fetch-mode', b'navigate')
    ▶ 08 = {tuple: 2} (b'sec-fetch-user', b'?1')
    ▶ 09 = {tuple: 2} (b'sec-fetch-dest', b'document')
    ▶ 10 = {tuple: 2} (b'accept-encoding', b'gzip, deflate, br')
    ▶ 11 = {tuple: 2} (b'accept-language', b'en-US,en;q=0.9,zh-CN;q=0.8,zh;q=0.7')
    ▶ 12 = {tuple: 2} (b'cookie', b'adminer_lang=en; adminer_settings=; adminer_engine=; admine... View
      __len__ = {int} 13
    __len__ = {int} 12
```

图 5-3 断点截图

分段传输大文件。

send也是一个异步函数，用于向前端发送信息，所发送的信息结构与从前端接收的信息结构类似。一个向前端发送简单信息的示例代码如下：

```
"""
第5章/asgi_app/simple_asgi.py
"""


async def app(scope, receive, send):
    # 向前端发送 HTTP 协议头,包括了 HTTP 状态与协议头
    await send({
        'type': 'http.response.start',
        'status': 200,
        'headers': [
            [b'content-type', b'text/html'],
        ]
    })
```

```
    #向前端发送数据,如果数据庞大,则可以分段发送
    await send({
        'type': 'http.response.body',
        'body': b"Hello World",
        'more_body': False
    })
```

除了常规数据通信外,ASGI 还规定了生命周期管理接口,可以用于侦听服务器的启动与关闭。在实际开发工作中,这非常有用,可以用来执行初始化工作与收尾工作,生命周期管理的运用代码如下:

```
"""
第 5 章/asgi_app/asgi_lc.py
"""


async def app(scope, receive, send):
    request_type = scope['type']
    if request_type == 'lifespan':
        while True:
            message = await receive()
            if message['type'] == 'lifespan.startup':
                await send({'type': 'lifespan.startup.complete'})
            elif message['type'] == 'lifespan.shutdown':
                await send({'type': 'lifespan.shutdown.complete'})
                break
```

当 scpoe['type']的类型是 lifespan 时,意味着该请求的类型是生命周期,该请求会在服务器启动之初发生,接下来应该实现对生命周期的管理。

通过无限循环不断侦听请求状态的变化,当读到 message['type']是 lifespan.startup 时执行初始化操作,在操作完成后向前端(协议层)发送 lifespan.startup.complete 信息,协议层可理解为服务器已经启动完成,可以正常接受浏览器请求了。

当读到 message['type']是 lifespan.shutdown 时,意味服务要关闭,可能是由于服务器管理员执行了关闭指令,那么在这里就需要执行收尾工作,例如释放相应资源等。在收尾完成后向协议层发送 lifespan.shutdown.complete 信息,表明此时协议层可以放心地关闭服务器。

一个完整的基于 ASGI 的 Hello World 示例代码如下:

```
"""
第 5 章/asgi_app/asgi.py
"""


async def app(scope, receive, send):
    request_type = scope['type']
    if request_type == 'http':
        await send({
```

```
                'type': 'http.response.start',
                'status': 200,
                'headers': [
                    [b'content-type', b'text/html'],
                ]
            })
            await send({
                'type': 'http.response.body',
                'body': b"Hello World",
                'more_body': False
            })
        elif request_type == 'lifespan':
            while True:
                message = await receive()
                if message['type'] == 'lifespan.startup':
                    await send({'type': 'lifespan.startup.complete'})
                elif message['type'] == 'lifespan.shutdown':
                    await send({'type': 'lifespan.shutdown.complete'})
                    break
        else:
            raise NotImplementedError()
```

5.3 Uvicorn

2min

Uvicorn是ASGI的一个协议层实现，一个轻量级的ASGI服务器，基于uvloop和httptools实现，运行速度极快。

uvloop是一个高效的基于异步IO的事件循环框架，底层实现由libuv承载。libuv是一个使用C语言开发的支持高并发的异步IO库，由Node.js的作者开发，作为Node.js的底层IO库实现，如今已经发展得相当成熟稳定。

要使用Uvicorn需要先通过命令pip install uvicorn安装该依赖项，项目结构如图5-4所示。

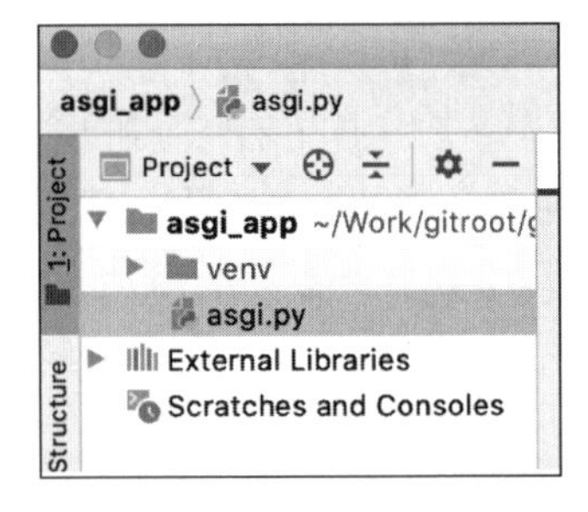

图5-4 ASGI项目文件结构

其中asgi.py文件即第5章/asgi_app/asgi.py。

接下来在终端输入uvicorn asgi:app以启动该服务器，效果如图5-5所示。

```
Terminal: Local × +
(venv) yunp@yunps-MBP asgi_app % uvicorn asgi:app
INFO:     Started server process [8574]
INFO:     Waiting for application startup.
INFO:     Application startup complete.
INFO:     Uvicorn running on http://127.0.0.1:8000 (Press CTRL+C to quit)
```

图5-5 启动ASGI服务器

服务器启动后，可使用浏览器通过 http://127.0.0.1:8000 访问该站点，结果如图 5-6 所示。

图 5-6 页面访问结果

为了向用户提供更加安全的服务，现代网站都需要支持 HTTPS，Uvicorn 也提供了对 HTTPS 的支持，使用起来也相当方便。

首先准备好 HTTPS 证书文件，如图 5-7 所示。

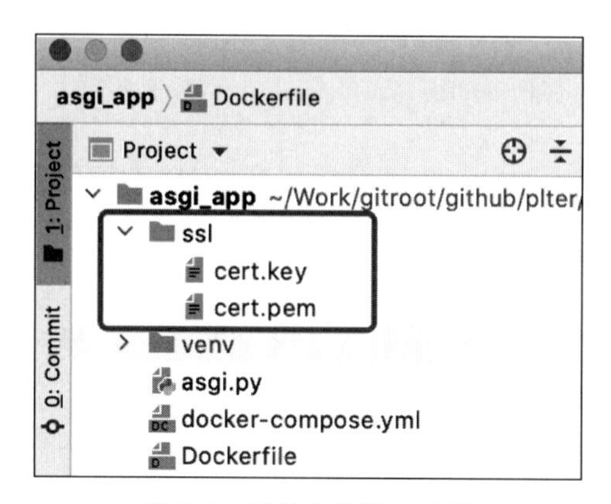

图 5-7 证书文件所在目录

接下来通过命令 uvicorn --ssl-certfile ./ssl/cert.pem --ssl-keyfile ./ssl/cert.key asgi:app 来启动该服务器，如图 5-8 所示。

```
Terminal:  Local ×  +
(venv) yunp@yunps-MBP asgi_app % uvicorn --ssl-certfile ./ssl/cert.pem --ssl-keyfile ./ssl/cert.key asgi:app
INFO:     Started server process [18505]
INFO:     Waiting for application startup.
INFO:     Application startup complete.
INFO:     Uvicorn running on https://127.0.0.1:8000 (Press CTRL+C to quit)
```

图 5-8 以 HTTPS 方式启动服务器

容器化对应的 Dockerfile 文件内容如下：

```
#第 5 章/asgi_app/Dockerfile
FROM python:3-slim

RUN pip3 install -i https://pypi.tuna.tsinghua.edu.cn/simple uvicorn
```

容器化对应的 docker-compose.yml 内容如下：

```
#第 5 章/asgi_app/docker-compose.yml
version: "3"

services:
  web:
    build: .
    restart: always
    tty: true
```

```
    ports:
      - "8000:8000"
    volumes:
      - ".:/opt/"
    working_dir: "/opt/"
    command: uvicorn --host 0.0.0.0 asgi:app
```

5.4 Daphne

Daphne 是一个功能强大的 ASGI 协议层实现，支持 HTTP、HTTP2 和 WebSocket 协议，由 Django 团队开发，用于支持 DjangoChannels（Channels 是一个由 Django 团队开发的可让 Django 支持 WebSocket、HTTP 长连接和其他异步编程的框架）的功能。

如果要使用 Daphne，需要先通过命令 pip install daphne 进行安装，然后通过命令 daphne asgi:app 启动服务器，如图 5-9 所示。

```
Terminal:  Local ×  +
(venv) yunp@yunps-MBP daphne_asgi_app % daphne asgi:app
2020-09-11 09:44:47,972 INFO     Starting server at tcp:port=8000:interface=127.0.0.1
2020-09-11 09:44:47,973 INFO     HTTP/2 support not enabled (install the http2 and tls Twisted extras)
2020-09-11 09:44:47,973 INFO     Configuring endpoint tcp:port=8000:interface=127.0.0.1
2020-09-11 09:44:47,975 INFO     Listening on TCP address 127.0.0.1:8000
```

图 5-9　启动 Daphne 服务器

此时可以使用浏览器通过 http://127.0.0.1:8000 进行访问，结果如图 5-10 所示。

Daphne 也支持 HTTPS，需要先准备好 HTTPS 证书，如图 5-11 所示。

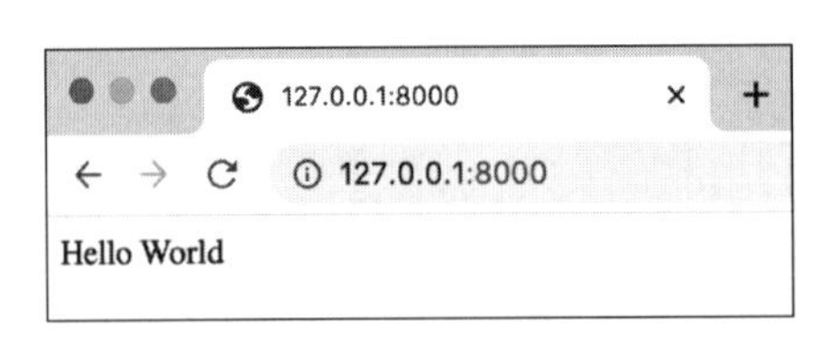

图 5-10　页面访问结果

图 5-11　证书文件所在目录

接下来用命令 daphne-e ssl:8443:privateKey=./ssl/cert.key:certKey=./ssl/cert.pem asgi:app 启动服务器，如图 5-12 所示。

```
daphne_asgi_app % daphne -e ssl:8443:privateKey=./ssl/cert.key:certKey=./ssl/cert.pem asgi:app
01 INFO     Starting server at ssl:8443:privateKey=./ssl/cert.key:certKey=./ssl/cert.pem
01 INFO     HTTP/2 support not enabled (install the http2 and tls Twisted extras)
01 INFO     Configuring endpoint ssl:8443:privateKey=./ssl/cert.key:certKey=./ssl/cert.pem
06 INFO     Listening on TCP address 0.0.0.0:8443
```

图 5-12　启动 Daphne HTTPS 服务器

此时可以使用浏览器通过 https://127.0.0.1:8443 进行访问,结果如图 5-13 所示。

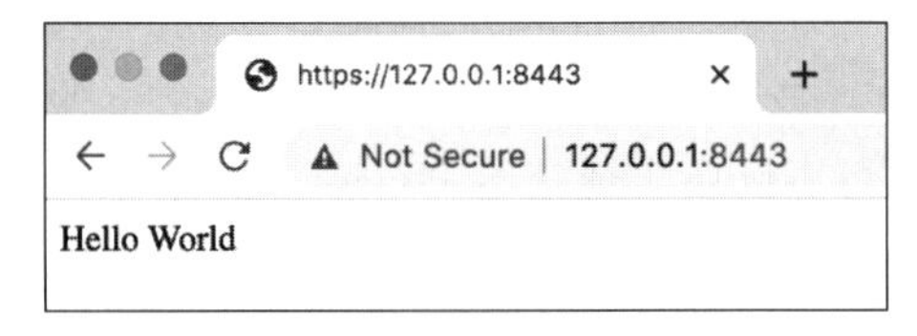

图 5-13 访问 HTTPS 页面

容器化对应的 Dockerfile 文件内容如下:

```
#第 5 章/daphne_asgi_app/Dockerfile
FROM python:3 - slim

#因为安装 Daphne 的过程中可能需要编译原生代码,所以需要安装 GCC 编译器
RUN apt update&&apt install  - y gcc make
RUN pip3 install  - i https://pypi.tuna.tsinghua.edu.cn/simple daphne
```

容器化对应的 docker-compose.yml 文件内容如下:

```
#第 5 章/daphne_asgi_app/docker - compose.yml
version: "3"

services:
  web:
    build: .
    restart: always
    tty: true
    ports:
      - "8000:8000"
    volumes:
      - ".:/opt/"
    working_dir: "/opt/"
    command: daphne  - b 0.0.0.0 asgi:app
```

5.5 Django 搭配 ASGI

Django 是一个非常强大的 Python Web 开发框架,也是目前 Python 语言使用最广泛的 Web 开发框架,很多人在入门 Python Web 开发中所学习的第一个技术框架就是 Django,此框架具有易学、易用、功能强大、文档优秀、技术支持完整等优点,这些优点使 Django 在近十几年来一直都是 Python 语言中最受欢迎的 Web 开发框架。

随着 ASGI 技术的快速发展,Django 率先对 ASGI 提供了支持,目前官方已经发行了支持 ASGI 的稳定版本框架。

与 AIOHTTP 不同,Django 对于技术的分层非常明确,Django 属于应用层实现,协议层可以自由切换。AIOHTTP 同时实现了协议层与应用层,并将所有的技术混合在一起,虽然使用起来相当方便，但是如果你期望后期随着技术的迁移和项目架构的改动而更换协议层进行实现，将会变得非常麻烦。而 Django 框架是单纯的应用层实现，无须担心协议层的技术实现，可以通过 WSGI 技术将协议层实现交给 Apache 服务器或者 Nginx 服务器实现，也可以通过 ASGI 技术将协议层实现交给 Uvicorn 或者 Daphne 等实现。

接下来我们一步一步来演示如何将 Django 部署在 ASGI 服务器上。

使用 PyCharm 创建一个新项目，名为 django_proj，并基于 Python 3.8 创建一个运行环境，如图 5-14 所示。

```
New Project
Location: /Users/yunp/Work/gitroot/github/plter/PythonAIOProgramming/django_proj
▼ Python Interpreter: New Virtualenv environment
◉ New environment using  Virtualenv
    Location: /Users/yunp/Work/gitroot/github/plter/PythonAIOProgramming/django_proj/venv
    Base interpreter: Python 3.8 /usr/local/bin/python3.8
    ☐ Inherit global site-packages
    ☐ Make available to all projects
○ Existing interpreter
    Interpreter: Python 3.8 /usr/local/bin/python3.8
☐ Create a main.py welcome script
  Create a Python script that provides an entry point to coding in PyCharm.
                                                    Cancel   Create
```

图 5-14　创建项目

项目创建完成后通过命令 pip install Django 安装 Django，如图 5-15 所示。

```
Terminal:  Local ×  +
(venv) yunp@yunps-MBP django_proj % pip install Django
Collecting Django
  Downloading Django-3.1.1-py3-none-any.whl (7.8 MB)
     |████████████████████████████████| 7.8 MB 89 kB/s
Collecting pytz
  Downloading pytz-2020.1-py2.py3-none-any.whl (510 kB)
     |████████████████████████████████| 510 kB 46 kB/s
Collecting asgiref~=3.2.10
  Using cached asgiref-3.2.10-py3-none-any.whl (19 kB)
Collecting sqlparse>=0.2.2
  Downloading sqlparse-0.3.1-py2.py3-none-any.whl (40 kB)
     |████████████████████████████████| 40 kB 53 kB/s
Installing collected packages: pytz, asgiref, sqlparse, Django
Successfully installed Django-3.1.1 asgiref-3.2.10 pytz-2020.1 sqlparse-0.3.1
```

图 5-15　安装 Django

通过命令 django-admin startproject web 创建一个 Django 项目，如图 5-16 所示。

通过命令 pip install uvicorn 安装 Uvicorn 环境，如图 5-17 所示。

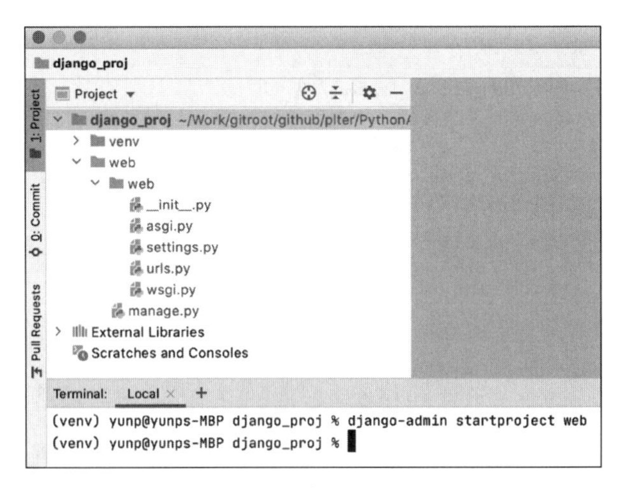

图 5-16　创建 Django 项目

```
Terminal:  Local
(venv) yunp@yunps-MBP django_proj % pip install uvicorn
Collecting uvicorn
  Downloading uvicorn-0.11.8-py3-none-any.whl (43 kB)
     |████████████████████████████████| 43 kB 124 kB/s
Collecting h11<0.10,>=0.8
  Downloading h11-0.9.0-py2.py3-none-any.whl (53 kB)
     |████████████████████████████████| 53 kB 72 kB/s
Collecting websockets==8.*
  Downloading websockets-8.1-cp38-cp38-macosx_10_9_x86_64.whl (64 kB)
     |████████████████████████████████| 64 kB 96 kB/s
Collecting uvloop>=0.14.0; sys_platform != "win32" and sys_platform != "cygwin" and platf
  Downloading uvloop-0.14.0-cp38-cp38-macosx_10_11_x86_64.whl (1.5 MB)
     |████████████████████████████████| 1.5 MB 59 kB/s
Collecting click==7.*
  Downloading click-7.1.2-py2.py3-none-any.whl (82 kB)
     |████████████████████████████████| 82 kB 111 kB/s
Collecting httptools==0.1.*; sys_platform != "win32" and sys_platform != "cygwin" and pla
  Downloading httptools-0.1.1-cp38-cp38-macosx_10_13_x86_64.whl (102 kB)
     |████████████████████████████████| 102 kB 68 kB/s
Installing collected packages: h11, websockets, uvloop, click, httptools, uvicorn
Successfully installed click-7.1.2 h11-0.9.0 httptools-0.1.1 uvicorn-0.11.8 uvloop-0.14.0
(venv) yunp@yunps-MBP django_proj %
```

图 5-17　安装 Uvicorn

使用命令 cd web 将工作目录切换到 web 目录，如图 5-18 所示。

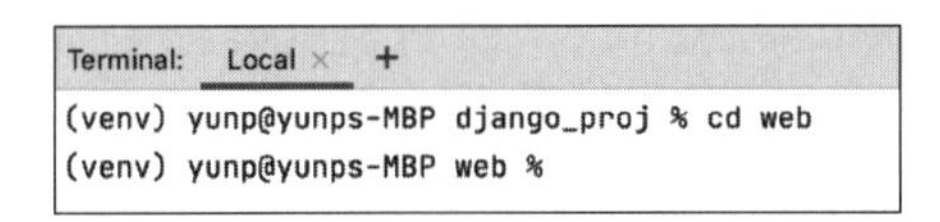

图 5-18　切换工作目录

使用命令 uvicorn web.asgi:application 启动服务器，如图 5-19 所示。

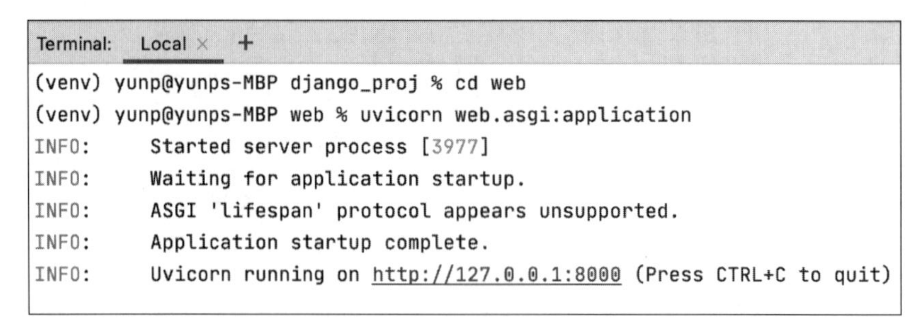

图 5-19 启动 Django 服务器

此时可以使用浏览器通过地址 http://127.0.0.1:8000 访问,效果如图 5-20 所示。

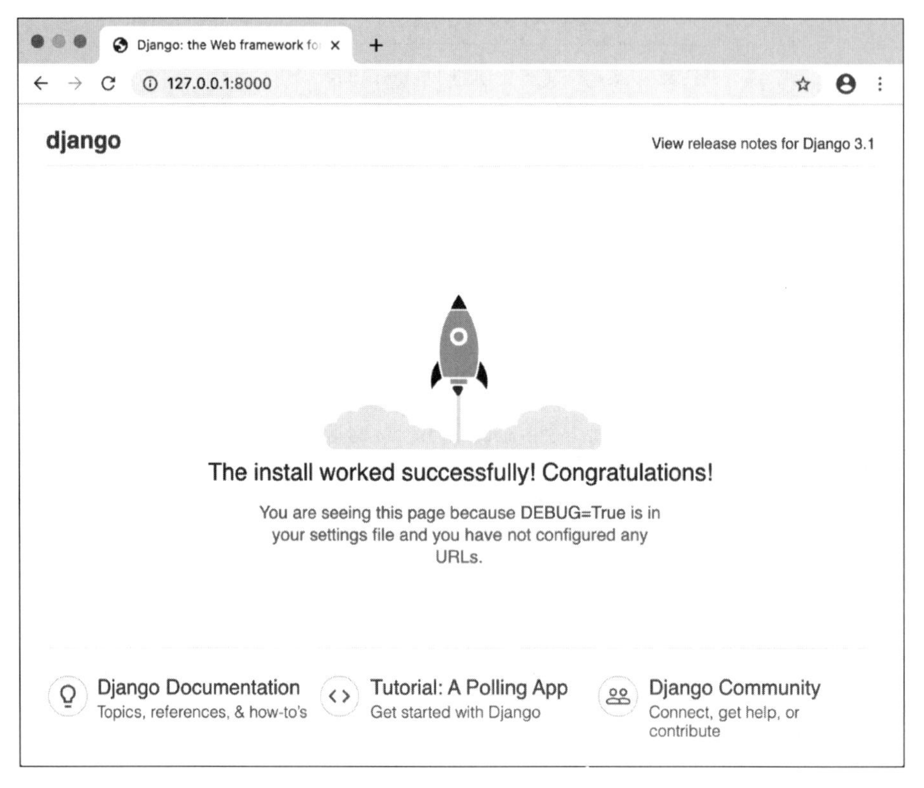

图 5-20 Django 站点首页

5.6 Quart

Quart 是一个基于 asyncio 的 Python Web 开发框架,提供了一种简单的在 Web 开发中使用 asyncio 的方式。

这里需要说明一下,为什么我们要了解这么多框架,因为目前 Python asyncio 生态正处于发展阶段,各种框架如雨后春笋般出现,呈现百家争鸣的状态。对于开发者来讲,不能押宝于任何一个框架,只能尝试学习更多的框架,才能在各种框架之间有一个实际的对比能力,在进行技术选型的时候能够更加准确,而不至于给项目后期带来不利影响。同时更应该去学习和掌握框架实现技术，从而拥有解决框架本身问题的能力，或者直接在项目中采用自己所开发的框架，以杜绝依赖第三方技术所带来的不稳定因素。

使用 Quart 之前需要先使用命令 pip install quart 安装该依赖项，之后创建一个文件名为 server.py 的文件，实现一个最简页面请求处理功能，代码如下：

```
"""
第 5 章/quart_app/server.py
"""
from quart import Quart

#创建一个 Quart 应用
app = Quart(__name__)


#处理对站点根路径的请求
@app.route('/')
async def hello():
    #向前端返回字符串
    return 'hello'

app.run()
```

通过命令 python server.py 启动该服务器，如图 5-21 所示。

```
Terminal:  Local ×  +
(venv) yunp@yunps-MBP quart_app % python server.py
Running on http://127.0.0.1:5000 (CTRL + C to quit)
[2020-09-11 23:01:28,790] Running on 127.0.0.1:5000 over http (CTRL + C to quit)
```

图 5-21 启动 Quart 服务器

此时可以通过地址 http://127.0.0.1:5000 访问该站点，如图 5-22 所示。

此时可以发现，Quart 是可以独立运行的，就像 AIOHTTP 一样，这是因为 Quart 的依赖项中有 Hypercorn（一个 ASGI 协议层实现，与 Uvicorn、Daphne 是同类型竞品，用法类似），在安装 Quart 时会自动安装 Hypercorn，而当使用 app.run 函数启动服务器时，会默认使用 Hypercorn 来启动服务器。

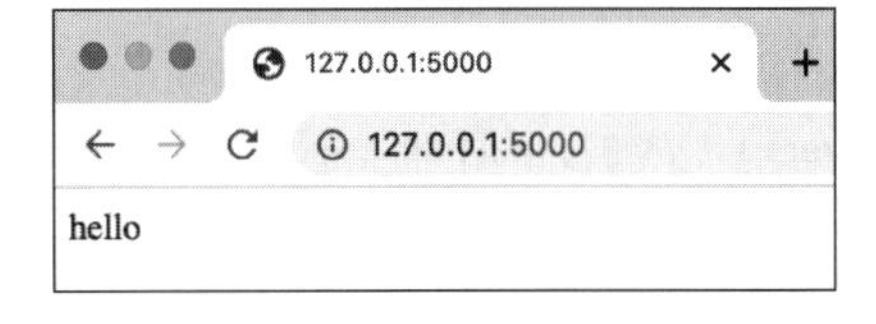

图 5-22 页面访问结果

也可以使用单独的 Hypercorn 去启动服务器，只需删除 app.run()这行代码，源码如下：

```
"""
第 5 章/quart_app/asgi.py
"""
```

```
from quart import Quart

# 创建一个 Quart 应用
app = Quart(__name__)


# 处理对站点根路径的请求
@app.route('/')
async def hello():
    # 向前端返回字符串
    return 'hello'
```

接下来通过命令 hypercorn server:app 启动服务器,如图 5-23 所示。

```
Terminal:   Local ×   +
(venv) yunp@yunps-MBP quart_app % hypercorn server:app
Running on 127.0.0.1:8000 over http (CTRL + C to quit)
```

图 5-23　使用 hypercorn 命令启动服务器

在安装 Uvicorn 之后也可以使用命令 uvicorn server:app 启动服务器,如图 5-24 所示。

```
Terminal:   Local ×   +
(venv) yunp@yunps-MBP quart_app % uvicorn server:app
INFO:     Started server process [5943]
INFO:     Waiting for application startup.
INFO:     Application startup complete.
INFO:     Uvicorn running on http://127.0.0.1:8000 (Press CTRL+C to quit)
```

图 5-24　使用 uvicorn 命令启动服务器

同样在安装 Daphne 之后也可以使用命令 daphne server:app 启动服务器,如图 5-25 所示。

```
Terminal:   Local ×   +
(venv) yunp@yunps-MBP quart_app % daphne server:app
2020-09-11 23:26:23,761 INFO     Starting server at tcp:port=8000:interface=127.0.0.1
2020-09-11 23:26:23,761 INFO     HTTP/2 support enabled
2020-09-11 23:26:23,761 INFO     Configuring endpoint tcp:port=8000:interface=127.0.0.1
2020-09-11 23:26:23,762 INFO     Listening on TCP address 127.0.0.1:8000
```

图 5-25　使用 daphne 命令启动服务器

5.7 Starlette

Starlette 是一个轻量级的高性能应用层框架,构建于 ASGI 之上,框架本身不带 ASGI 协议层,所以运行时需要手动安装协议层依赖项,例如 uvicorn。

要使用 Starlette,需要先用命令 pip install starlette 安装该框架,如图 5-26 所示。

```
Terminal:  Local ×  +
(venv) yunp@yunps-MBP starlette_app % pip install starlette
Requirement already satisfied: starlette in ./venv/lib/python3.8/site-packages (0.13.8)
(venv) yunp@yunps-MBP starlette_app %
```

图 5-26 安装 Starlette

创建一个名为 server.py 的文件，在其中输入 Hello World 示例代码如下：

```python
"""
第 5 章/starlette_app/server.py
"""
from starlette.applications import Starlette
from starlette.responses import HTMLResponse
from starlette.routing import Route


async def homepage(request):
    return HTMLResponse("Hello World")


app = Starlette(debug=True, routes=[
    Route('/', homepage),
])
```

若要运行该服务器，需要先使用命令 pip install uvicorn 安装 Uvicorn，也可以尝试使用其他的协议层实现，如 Daphne 或者 Hypercorn。在安装 Uvicorn 之后需要使用命令 uvicorn server:app 启动服务器，如图 5-27 所示。

```
Terminal:  Local ×  +
(venv) yunp@yunps-MBP starlette_app % ucivorn server:app
zsh: command not found: ucivorn
(venv) yunp@yunps-MBP starlette_app % uvicorn server:app
INFO:     Started server process [9962]
INFO:     Waiting for application startup.
INFO:     Application startup complete.
INFO:     Uvicorn running on http://127.0.0.1:8000 (Press CTRL+C to quit)
```

图 5-27 启动 Starlette 服务器

第 6 章

Tornado

Tornado 是一个 Python Web 开发框架，也是一个异步网络库。它是一个使用广泛、经得起考验的老牌异步网络库，早在 asyncio 库之前就已经形成了完整的异步 IO 编程体系，其所提供的功能也相当稳定及强大，因此被广泛应用于游戏服务器、网站服务器的开发。

4min

6.1 TCP 服务器

在使用 Tornado 之前，需要先使用命令 pip install tornado 安装该库，如图 6-1 所示。

```
Terminal:  Local ×  +
(venv) yunp@yunps-MBP tornado_socket_server % pip install tornado
Collecting tornado
  Using cached tornado-6.0.4.tar.gz (496 kB)
Using legacy 'setup.py install' for tornado, since package 'wheel' is not installed.
Installing collected packages: tornado
    Running setup.py install for tornado ... done
Successfully installed tornado-6.0.4
```

图 6-1　安装 Tornado

创建一个 tornado.tcpserver.TCPServer 的子类，用于处理连接请求，代码如下：

```
class EchoServer(TCPServer):
    async def handle_stream(self, stream, address):
        pass
```

接下来创建该服务器实例，并侦听一个端口，代码如下：

```
server = EchoServer()
server.listen(8888)
```

最后启动 IO 事件循环，代码如下：

```
IOLoop.current().start()
```

以这个流程操作便可启动一个 Tornado Socket 服务器，具体的业务逻辑可以通过重写 TCPServer 的 handle_stream 函数实现，接下来以一个向连接终端发送 6 条数据的业务逻辑为例演示最简流程，代码如下：

```
"""
第 6 章/tornado_tcp_server/tornado_tcp_server.py
"""

import asyncio, time

from tornado.iostream import StreamClosedError
from tornado.tcpserver import TCPServer
from tornado.ioloop import IOLoop


class EchoServer(TCPServer):
    async def handle_stream(self, stream, address):
        for i in range(6):
            try:
                #向连接终端发送数据
                await stream.write(
                    f"[{time.strftime('%X')}] Count {i}\n".encode("utf-8")
                )
                #等待 1s
                await asyncio.sleep(1)
            except StreamClosedError:
                break


server = EchoServer()
server.listen(8888)
try:
    IOLoop.current().start()
except KeyboardInterrupt as e:
    print("Exit")
```

在启动该服务器后，使用 telnet 127.0.0.1 8888 命令连接，效果如图 6-2 所示。

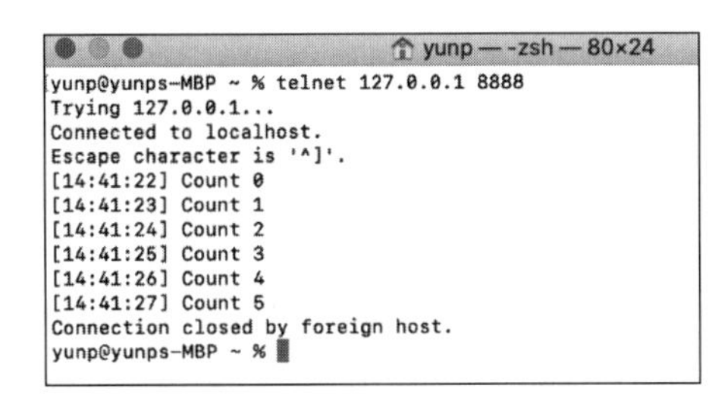

图 6-2 连接 TCP 服务器效果

6.2 HTTP 服务器

Tornado 是一个强大的 Python Web 开发框架，功能完整而且使用简单，在使用 Tornado 之前需要先使用命令 pip install tornado 进行安装，然后只需几行代码便可创建一

个 Web 服务器。代码如下：

```
"""
第 6 章/tornado_http_server/tornado_http_server1.py
"""

import tornado.ioloop
import tornado.web

if __name__ == '__main__':
    #创建 Web 服务器应用
    app = tornado.web.Application()
    app.listen(8888) #侦听端口 8888
    tornado.ioloop.IOLoop.current().start() #启动
```

启动该服务器，用浏览器访问效果如图 6-3 所示。

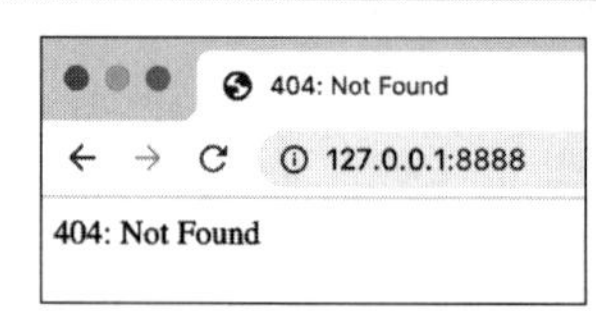

图 6-3　空服务器运行效果

可以看到，服务器已经能够处理请求了，但是还访问不到主页，这是因为主页暂未进行配置，可通过代码配置主页，代码如下：

```
"""
第 6 章/tornado_http_server/tornado_http_server2.py
"""

import tornado.ioloop
import tornado.web


#HomePage 是一个继承自 tornado.web.RequestHandler 的类
class HomePage(tornado.web.RequestHandler):
    def get(self):
        self.write("Hello World")


if __name__ == '__main__':
    #创建 Web 服务器应用
    app = tornado.web.Application([
        #配置一个请求处理器并命名为 HomePage,然后将其映射到网站根路径"/"上,
        #而 HomePage 是一个继承自 tornado.web.RequestHandler 的类
        ("/", HomePage)
    ])
    app.listen(8888) #侦听端口 8888
    tornado.ioloop.IOLoop.current().start() #启动
```

图 6-4　配置根页面的运行效果

运行该程序，则使用浏览器访问的效果如图 6-4 所示。

如果希望处理 post 请求，则需要重写 tornado.web.RequestHandler 类的 post 函数，如下所示：

```
class HomePage(tornado.web.RequestHandler):

    def post(self):
        self.write("Handle post request")
```

如果希望在程序中执行异步代码，则可为处理函数添加 async 修饰，如下所示：

```
class HomePage(tornado.web.RequestHandler):
    async def get(self):
        self.write("Hello World")
```

6.3 路由

路由功能用于将路径与处理程序进行对应，Tornado 拥有强大的路由功能，首先来认识最简单的路由配置，代码如下：

```
app = tornado.web.Application([
    ("/", HomePage), #将 HomePage 映射到/根路径上
    ("/user", Users) #将 Users 映射到/users 路径上
])
```

Tornado 的路由不仅可以使用简单字符串进行配置，还可以使用正则表达式进行配置，代码如下：

```
(r"/app.*", AppPage)
```

这种写法在实际开发中非常有用，可以将所有以/app 开头的请求均映射到 AppPage 程序进行处理，对于需要共同初始化代码的程序来讲，可以免去大量重复性的工作。完整代码如下：

```
"""
第 6 章/tornado_routes/routes.py
"""
import tornado.ioloop
import tornado.web
import tornado.routing


#HomePage 是一个继承自 tornado.web.RequestHandler 的类
class HomePage(tornado.web.RequestHandler):
    def get(self):
        self.write("Home")


class Users(tornado.web.RequestHandler):
    def get(self):
```

```
            self.write("Users")


class AppPage(tornado.web.RequestHandler):
    def get(self):
        self.write(f"Request path is: {self.request.path}")


if __name__ == '__main__':
    # 创建 Web 服务器应用
    app = tornado.web.Application([
        ("/", HomePage), # 将 HomePage 映射到/根路径上
        ("/user", Users), # 将 Users 映射到/users 路径上
        (r"/app.*", AppPage)
    ])
    app.listen(8888) # 侦听端口 8888
    tornado.ioloop.IOLoop.current().start() # 启动
```

访问 AppPage 的效果如图 6-5 所示。

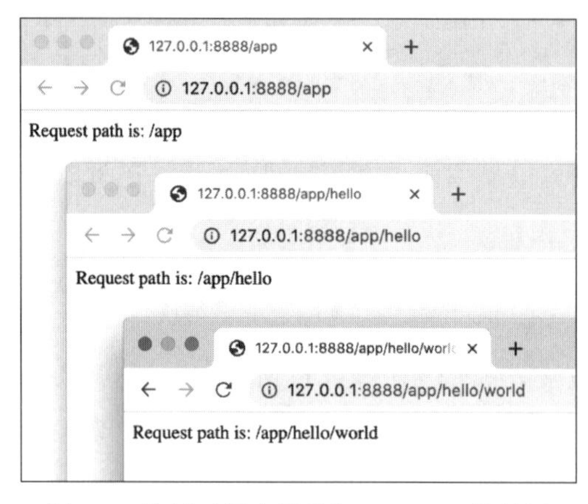

图 6-5 使用不同路径访问 AppPage 效果图

如果要截取路径中的一部分作为参数使用，也可以使用正则表达式的截取变量语法，当然对应的处理函数需要具有与配置路径的正则表达式匹配的接收参数的功能，代码如下：

```
"""
第 6 章/tornado_routes/routes_rapp.py
"""
import tornado.ioloop
import tornado.web
import tornado.routing


class AppPage(tornado.web.RequestHandler):
    def get(self, path_arg1, path_arg2):
```

```
        """
该函数必须可以接收两个参数
        :param path_arg1: 对应正则表达式截取的第 1 个变量
        :param path_arg2: 对应正则表达式截取的第 2 个变量
        :return:
        """
        self.write(f"Arg1 is: {path_arg1}, arg2 is:{path_arg2}")


if __name__ == '__main__':
    # 创建 Web 服务器应用
    app = tornado.web.Application([
        # 该路径设置了两个截取的变量
        (r"/app/(.*)/(.*)", AppPage)
    ])
    app.listen(8888) # 侦听端口 8888
    tornado.ioloop.IOLoop.current().start() # 启动
```

使用浏览器访问该服务器的效果如图 6-6 所示。

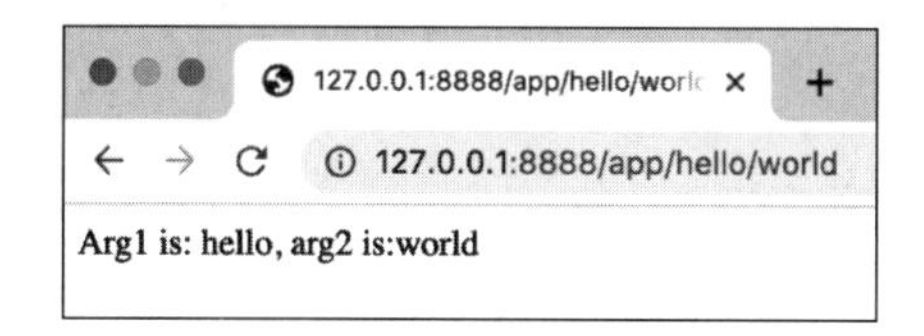

图 6-6 路径参数效果

Tornado 还支持多个应用运行于同一台服务器上，代码如下：

```
"""
第 6 章/tornado_routes/multi_app.py
"""
import tornado.ioloop
from tornado.web import Application
from tornado.routing import Rule, RuleRouter, PathMatches


class HandlerInApp1(tornado.web.RequestHandler):
    def get(self):
        self.write("Handler in app1")


app1 = Application([
    (r"/app1/handler", HandlerInApp1)
])


class HandlerInApp2(tornado.web.RequestHandler):
    def get(self):
```

```
        self.write("Handler in app2")


app2 = Application([
    (r"/app2/handler", HandlerInApp2)
])

if __name__ == '__main__':
    #创建服务器
    server = tornado.web.HTTPServer(
        RuleRouter([
            Rule(PathMatches(r"/app1.*"), app1),
            Rule(PathMatches(r"/app2.*"), app2)
        ])
    )
    #侦听端口 8888
    server.listen(8888)
    #启动
    try:
        tornado.ioloop.IOLoop.current().start()
    except KeyboardInterrupt:
        print("Server stopped by user")
```

使用浏览器访问的效果如图 6-7 所示。

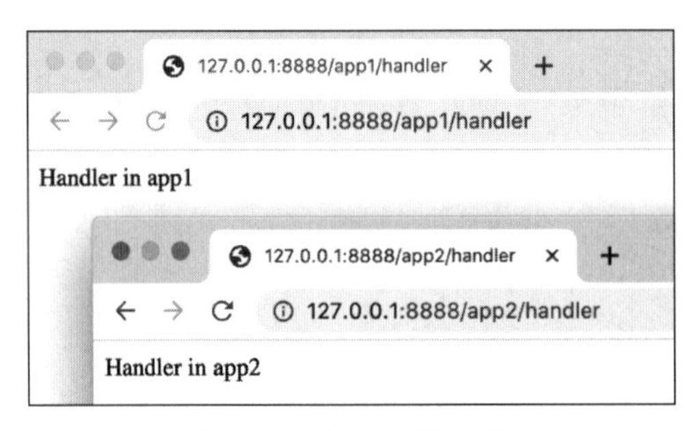

图 6-7 多应用服务器

6.4 处理静态文件

Tornado 提供了静态文件处理功能，虽然一般来说在实际部署的服务器上会采用 Apache 或者 Nginx 作为静态文件的服务器使用，但是 Tornado 自带的静态文件处理功能在开发阶段非常有用。因为 Tornado 的静态文件处理功能采用异步 IO 实现，理论上讲可以支持高并发，所以在中小型网站中，也可以直接使用 Tornado 来服务静态文件请求。

只需将处理程序设置为 tornado.web.StaticFileHandler 便可实现对静态文件进行处理，代码如下：

```
"""
第 6 章/tornado_static/tornado_static_server.py
"""

from tornado.web import Application, StaticFileHandler
from tornado.ioloop import IOLoop
import os

APP_ROOT = os.path.dirname(__file__)

if __name__ == '__main__':
    app = Application([
        (
            r"/static/(.*)",
            StaticFileHandler,
            #配置当前文件同目录下的 static 目录为静态文件所在目录
            dict(path=os.path.join(APP_ROOT, "static"))
        )
    ])
    #侦听端口 8888
    app.listen(8888)
    try:
        #启动
        IOLoop.current().start()
    except KeyboardInterrupt:
        print("Server stopped by user")
```

该项目的目录结构如图 6-8 所示。

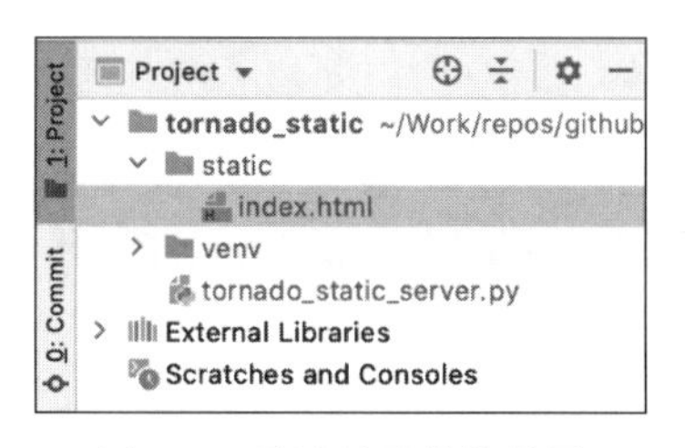

图 6-8 项目目录结构截图

其中 index.html 文件代码如下：

```
<!-- 第 6 章/tornado_static/static/index.html -->
<!DOCTYPE html>
<html lang="en">
<head>
<meta charset="UTF-8">
<title>标题</title>
</head>
<body>
这是一个静态文件
```

```
</body>
</html>
```

启动服务器后在浏览器中访问 http://127.0.0.1:8888/static/index.html，效果如图 6-9 所示。

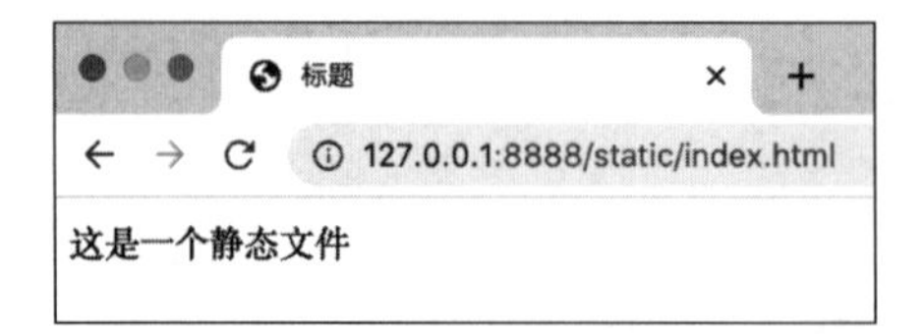

图 6-9 静态页面访问

虽然 Tornado 有强大的静态文件处理功能，但是相比 Apache 和 Nginx 这种专业的 HTTP 服务器来说还是弱一些，而且功能过于单一，而 Apache 和 Nginx 有完整的 HTTP 服务器功能，在实际部署的项目里，通常会使用 Apache 或 Nginx 作为网关。

接下来用 Apache 服务器处理静态文件请求以实现负载均衡，原理是将 /static 目录放在 Apache 的文档目录下，同时将除 /static 之外的其他所有请求代理到 Python 服务器，这样可以实现动态请求由 Python 处理，而静态请求由 Apache 处理。

首先需要启用 Apache 的 mod_proxy 与 mod_proxy_http 模块，配置代码如下：

```
#启用代理模块
LoadModule proxy_module modules/mod_proxy.so
#启用 http 代理
LoadModule proxy_http_module modules/mod_proxy_http.so
```

然后配置代理，代码如下：

```
#不代理 /static 目录，而是直接由 Apache 服务器处理该请求
ProxyPass /static/ !
ProxyPassReverse /static/ !

#将所有请求代理到 Python 服务器
ProxyPass / http://web:8888/
ProxyPassReverse / http://web:8888/
```

为了部署方便，此处还是使用 Docker 容器化技术实现该负载均衡方案，完整的 httpd.conf 文件内容如下：

```
#第 6 章/tornado_static/RunTime/gateway/httpd.conf
#

ServerRoot "/usr/local/apache2"
Listen 80

LoadModule mpm_event_module modules/mod_mpm_event.so
```

```
LoadModule authn_file_module modules/mod_authn_file.so
LoadModule authn_core_module modules/mod_authn_core.so
LoadModule authz_host_module modules/mod_authz_host.so
LoadModule authz_groupfile_module modules/mod_authz_groupfile.so
LoadModule authz_user_module modules/mod_authz_user.so
LoadModule authz_core_module modules/mod_authz_core.so
LoadModule access_compat_module modules/mod_access_compat.so
LoadModule auth_basic_module modules/mod_auth_basic.so
LoadModule reqtimeout_module modules/mod_reqtimeout.so
LoadModule filter_module modules/mod_filter.so
LoadModule mime_module modules/mod_mime.so
LoadModule log_config_module modules/mod_log_config.so
LoadModule env_module modules/mod_env.so
LoadModule headers_module modules/mod_headers.so
LoadModule setenvif_module modules/mod_setenvif.so
LoadModule version_module modules/mod_version.so

#启用代理模块
LoadModule proxy_module modules/mod_proxy.so
#启用 http 代理
LoadModule proxy_http_module modules/mod_proxy_http.so

LoadModule UNIXd_module modules/mod_UNIXd.so
LoadModule status_module modules/mod_status.so
LoadModule autoindex_module modules/mod_autoindex.so
<IfModule !mpm_prefork_module>
    #LoadModule cgid_module modules/mod_cgid.so
</IfModule>
<IfModule mpm_prefork_module>
    #LoadModule cgi_module modules/mod_cgi.so
</IfModule>
LoadModule dir_module modules/mod_dir.so
LoadModule alias_module modules/mod_alias.so

<IfModule UNIXd_module>
User daemon
Group daemon
</IfModule>

ServerAdmin you@example.com

<Directory />
    AllowOverride none
    Require all denied
</Directory>

DocumentRoot "/usr/local/apache2/htdocs"
<Directory "/usr/local/apache2/htdocs">
    Options Indexes FollowSymLinks
    AllowOverride None
```

```
    Require all granted
</Directory>

<IfModule dir_module>
    DirectoryIndex index.html
</IfModule>

<Files ".ht*">
    Require all denied
</Files>

ErrorLog /proc/self/fd/2

LogLevel warn

<IfModule log_config_module>
    LogFormat "%h %l %u %t \"%r\" %>s %b \"%{Referer}i\" \"%{User-Agent}i\"" combined
    LogFormat "%h %l %u %t \"%r\" %>s %b" common

<IfModule logio_module>
      # You need to enable mod_logio.c to use %I and %O
      LogFormat "%h %l %u %t \"%r\" %>s %b \"%{Referer}i\" \"%{User-Agent}i\" %I %O" combinedio
</IfModule>

    CustomLog /proc/self/fd/1 common
</IfModule>

<IfModule alias_module>
    ScriptAlias /cgi-bin/ "/usr/local/apache2/cgi-bin/"
</IfModule>

<IfModule cgid_module>
</IfModule>

<Directory "/usr/local/apache2/cgi-bin">
    AllowOverride None
    Options None
    Require all granted
</Directory>

<IfModule headers_module>
    RequestHeader unset Proxy early
</IfModule>

<IfModule mime_module>
    TypesConfig conf/mime.types
    AddType application/x-compress .Z
```

```
    AddType application/x-gzip .gz .tgz
</IfModule>

<IfModule proxy_html_module>
Include conf/extra/proxy-html.conf
</IfModule>

<IfModule ssl_module>
SSLRandomSeed startup builtin
SSLRandomSeed connect builtin
</IfModule>

#不代理 /static 目录,而是直接由 Apache 服务器处理该请求
ProxyPass /static/ !
ProxyPassReverse /static/ !

#将所有请求代理到 Python 服务器
ProxyPass / http://web:8888/
ProxyPassReverse / http://web:8888/
```

项目目录结构如图 6-10 所示。

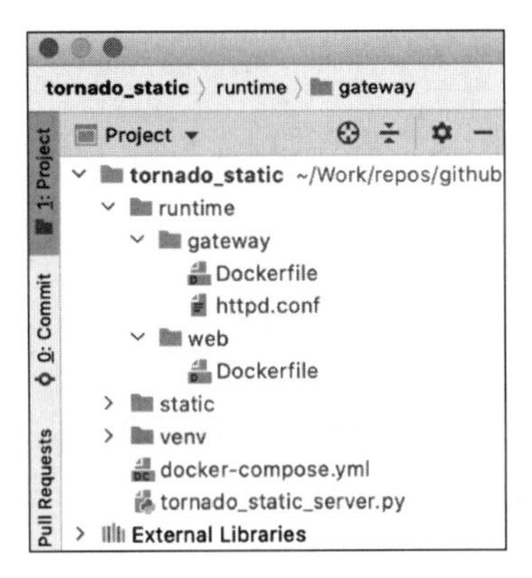

图 6-10 项目目录结构

docker-compose.yml 文件内容如下：

```
#第 6 章/tornado_static/docker-compose.yml

version: "3"

services:
  gateway:
    build: ./RunTime/gateway
    volumes:
      #将 static 目录映射到 Apache 服务器的文档目录下
      - "./static:/usr/local/apache2/htdocs/static"
    ports:
```

```
      #将 gateway 的 80 端口映射到本机 80 端口
      - 80:80

  web:
    build: ./RunTime/web
    volumes:
      #将当前目录映射到 Web 服务中的 /opt/web 目录
      - ".:/opt/web"
    #设置工作目录为 /opt/web
    working_dir: "/opt/web"
    #容器启动后执行该命令用于启动 Python 服务器
    command: python3 ./tornado_static_server.py
```

网关服务的 Dockerfile 文件内容如下：

```
#第 6 章/tornado_static/RunTime/gateway/Dockerfile

#基于 Apache 2.4 镜像
FROM httpd:2.4

#用自定义的配置文件替换该镜像的默认配置文件
COPY httpd.conf /usr/local/apache2/conf
```

Python 服务器的 Dockerfile 文件内容如下：

```
#第 6 章/tornado_static/RunTime/web/Dockerfile

#基于 python:3-slim 镜像构建 Web 服务镜像
FROM python:3-slim

#安装 tornado
RUN pip3 install -i https://pypi.tuna.tsinghua.edu.cn/simple tornado
```

配置完成后使用命令 docker-compose up -d 以启动所有服务，如图 6-11 所示。

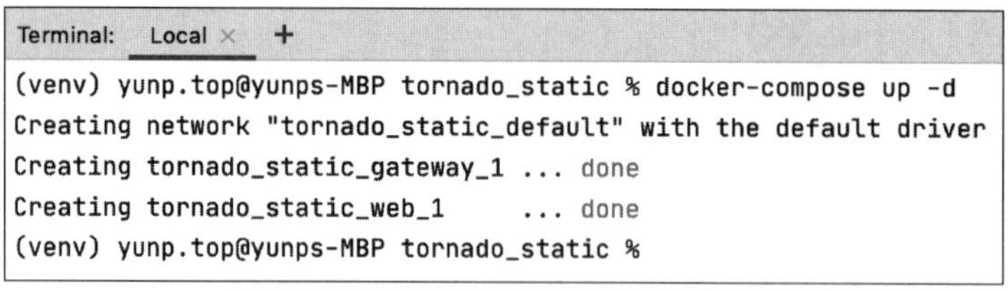

```
Terminal:  Local ×  +
(venv) yunp.top@yunps-MBP tornado_static % docker-compose up -d
Creating network "tornado_static_default" with the default driver
Creating tornado_static_gateway_1 ... done
Creating tornado_static_web_1     ... done
(venv) yunp.top@yunps-MBP tornado_static %
```

图 6-11 启动 Docker 服务

通过浏览器访问 http://127.0.0.1 可以看到该请求由 Python 服务器处理，如图 6-12 所示。

通过浏览器访问 http://127.0.0.1/static/index.html 可以看到该请求由 Apache 服务器处理，如图 6-13 所示。

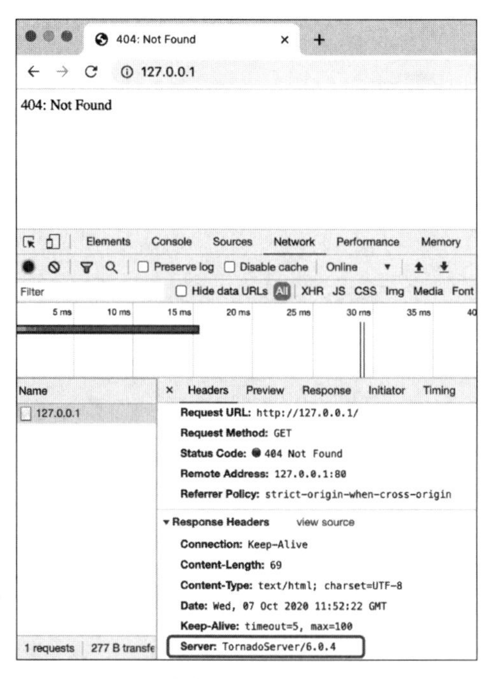

图 6-12　请求由 Python 服务器处理

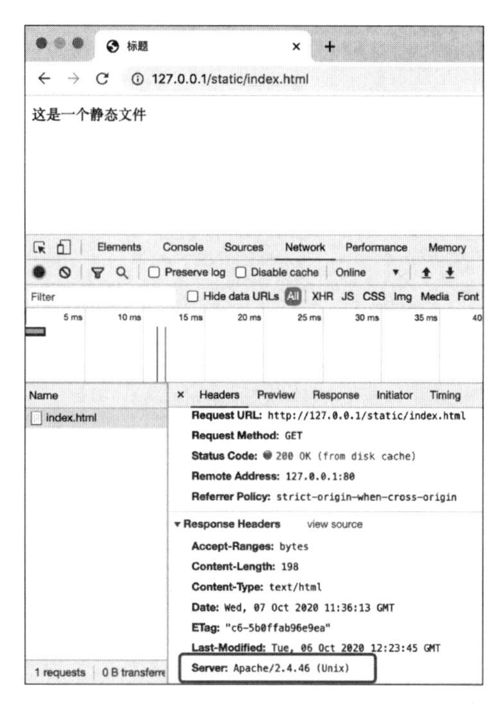

图 6-13　请求由 Apache 服务器处理

6.5 模板渲染

模板渲染功能大大提高了拼装页面的效率，在 Tornado 中内置了模板功能，使用起来极为方便。只需继承自 tornado.web.RequestHandler 类，并在内部调用其 render 函数便可实现模板渲染，代码如下：

```
class IndexHandler(tornado.web.RequestHandler):
    async def get(self):
        await self.render("index.html")
```

示例项目文件结构如图 6-14 所示。

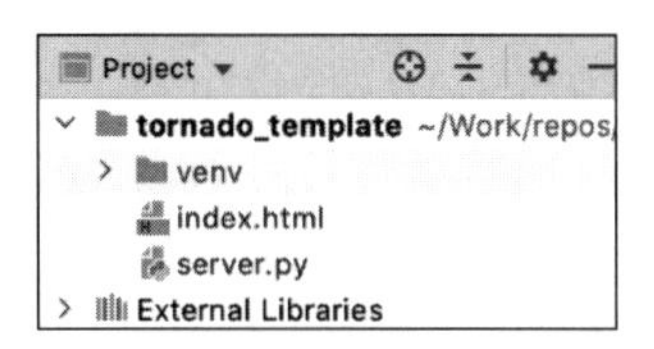

图 6-14　模板示例项目文件结构

其中 server.py 文件代码如下：

```
"""
第 6 章/tornado_template/server.py
"""
```

```
import tornado.ioloop
import tornado.web


class IndexHandler(tornado.web.RequestHandler):
    async def get(self):
        await self.render("index.html")


if __name__ == "__main__":
    application = tornado.web.Application([
        (r"/", IndexHandler),
    ])
    application.listen(8888)
    try:
        tornado.ioloop.IOLoop.current().start()
    except KeyboardInterrupt as e:
        print("User stopped the server")
```

index.html 文件代码如下：

```
<!-- 第6章/tornado_template/index.html -->
<!DOCTYPE html>
<html lang="en">
<head>
<meta charset="UTF-8">
<title>Title</title>
</head>
<body>
Default page
</body>
</html>
```

启动该服务器，在浏览器中通过地址 http://127.0.0.1:8888 访问，效果如图 6-15 所示。

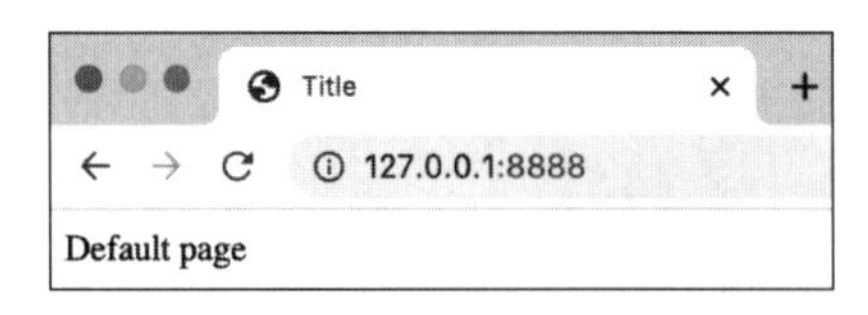

图 6-15 访问服务器效果

在实际开发工作中，不建议把所有种类的文件放在同一个目录，例如模板文件应当放在一个单独的目录中，如果需要将模板文件放在名为 template 的目录下，则可以通过应用配置(Application settings)实现，代码如下：

```
application = tornado.web.Application([
    (r"/", IndexHandler),
], template_path="template")
```

将文件结构调整为如图 6-16 所示。

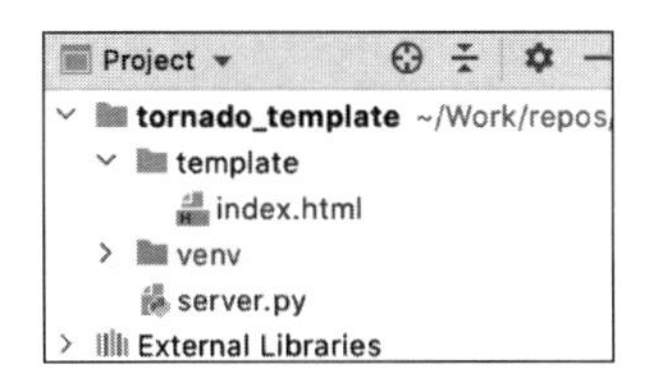

图 6-16 调整后的文件结构

与图 6-16 所对应的 server1.py 文件代码如下：

```
"""
第 6 章/tornado_template/server1.py
"""

import tornado.ioloop
import tornado.web
import os

SERVER_ROOT = os.path.dirname(__file__)


class IndexHandler(tornado.web.RequestHandler):
    async def get(self):
        await self.render("index.html")


if __name__ == "__main__":
    application = tornado.web.Application([
        (r"/", IndexHandler),
    ], template_path=os.path.join(SERVER_ROOT, "template"))
    application.listen(8888)
    try:
        tornado.ioloop.IOLoop.current().start()
    except KeyboardInterrupt as e:
        print("User stopped the server")
```

Tornado 内置的模板功能简单实用，支持继承、循环、流程控制等模板常用功能，其语法与 Jinja2 类似，但有细微差别。在 Jinja2 中，每个语句必须有对应结尾语句，例如 if 语句必须以 endif 结尾、block 语句必须以 endblock 结尾、for 语句必须以 endfor 结尾等，但在 Tornado 模板中，所有语句均由 end 结尾。

在 Tornado 中一个 block 的声明代码如下：

```
{% block body %}{% end %}
```

接下来介绍 Tornado 模板的继承功能，项目文件结构如图 6-17 所示。

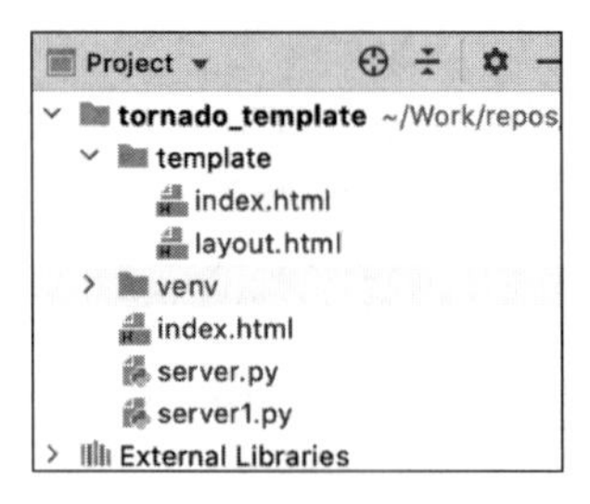

图 6-17　模板继承示例文件结构

其中 template/index.html 代码如下：

```
<!-- 第 6 章/tornado_template/template/index.html -->
{% extends "layout.html" %}

{% block body %} Hello World {% end %}
```

其中 template/layout.html 代码如下：

```
<!-- 第 6 章/tornado_template/template/layout.html -->
<!DOCTYPE html>
<html lang="en">
<head>
<meta charset="UTF-8">
<title>Title</title>
</head>
<body>
{% block body %}{% end %}
</body>
</html>
```

最终渲染出的界面效果如图 6-18 所示。

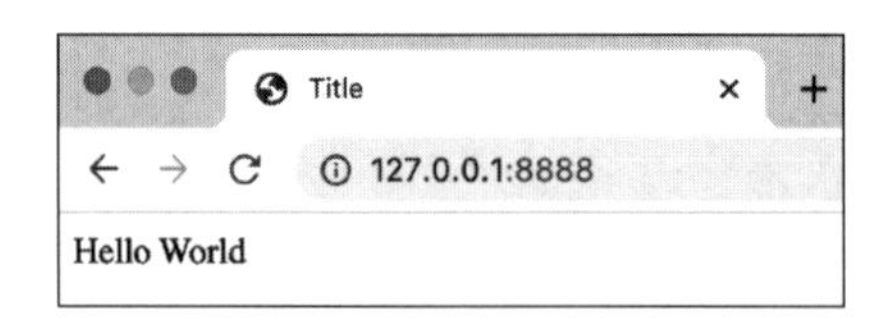

图 6-18　页面继承渲染效果

下面介绍模板中循环语句的使用方式，将 server1.py 复制为 server2.py，并对其中的代码进行修改，修改后的代码如下：

```
"""
第 6 章/tornado_template/server2.py
"""

import tornado.ioloop
import tornado.web
```

```
import os

SERVER_ROOT = os.path.dirname(__file__)


class IndexHandler(tornado.web.RequestHandler):
    async def get(self):
        await self.render("server2_index.html", users=['李明', '张亮'])


if __name__ == "__main__":
    application = tornado.web.Application([
        (r"/", IndexHandler),
    ], template_path=os.path.join(SERVER_ROOT, "template"))
    application.listen(8888)
    try:
        tornado.ioloop.IOLoop.current().start()
    except KeyboardInterrupt as e:
        print("User stopped the server")
```

与之对应的 server2_index.html 文件代码如下：

```
<!-- 第 6 章/tornado_template/template/server2_index.html -->
{% extends "layout.html" %}

{% block body %}
<div>
    Users:
<ul>
        {% for u in users %}
<li>{{ u }}</li>
        {% end %}
</ul>
</div>
{% end %}
```

最终渲染出的页面显示效果如图 6-19 所示。

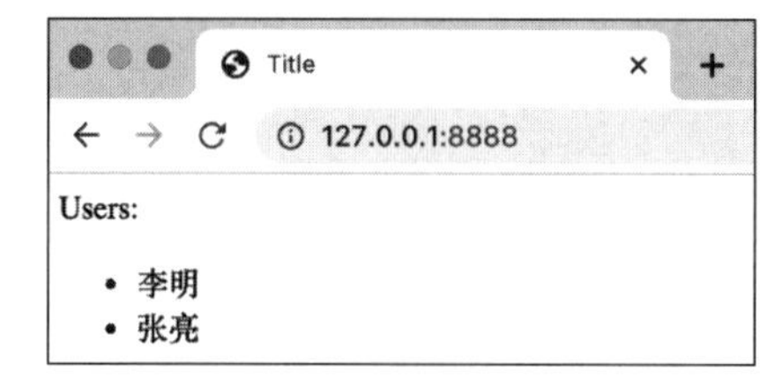

图 6-19　列表渲染效果

6.6 多语言支持

Tornado 提供了一种可以自动根据浏览器语言进行翻译的机制，要使用该机制，须先通过 tornado.locale.load_translations 函数加载语言表。

示例项目文件结构如图 6-20 所示。

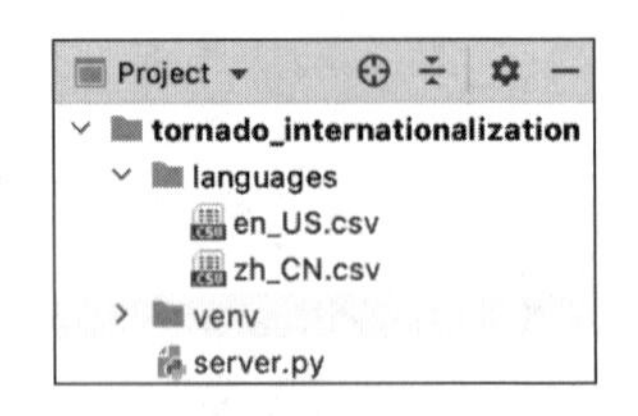

图 6-20 多语言项目文件结构

在该项目中，约定将所有的自然语言文件放在 languages 目录中，server.py 文件代码如下：

```
"""
第 6 章/tornado_internationalization/server.py
"""
import tornado.ioloop
import tornado.web
import tornado.locale
import os

SERVER_ROOT = os.path.dirname(__file__)


class MainHandler(tornado.web.RequestHandler):
    def get(self):
        #根据浏览器语言自动翻译
        words = self.locale.translate("Hello")
        self.write(f"{words}")

if __name__ == "__main__":
    #从指定的目录中加载语言表
    tornado.locale.load_translations(os.path.join(SERVER_ROOT, "languages"))
    application = tornado.web.Application([
        (r"/", MainHandler),
    ])
    application.listen(8888)
    try:
        tornado.ioloop.IOLoop.current().start()
    except KeyboardInterrupt:
        print("User stopped server")
```

en_US.csv 文件内容如下：

```
Hello,Hello
```

zh_CN.csv 文件内容如下：

```
Hello,你好
```

可以看出，这两个文件分别针对美国英语和简体中文进行翻译。

这里需要注意的是文件名命名规则是语言和地区中间的分割符是“_”而不是“-”，但浏览器请求时传递的语言字符串中间的分割符是“-”，如图 6-21 所示。

图 6-21 浏览器请求头中的语言字符串

在英文浏览器中访问服务器的效果如图 6-22 所示。

在中文浏览器中访问服务器的效果如图 6-23 所示。

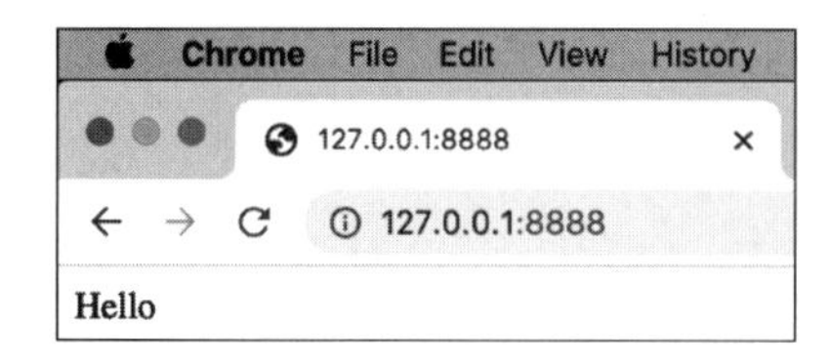

图 6-22 英文浏览器访问效果

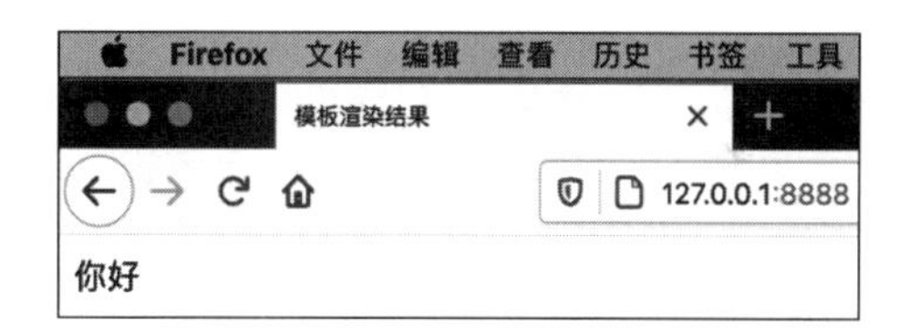

图 6-23 中文浏览器访问效果

接下来介绍如何在模板中使用翻译功能，将 server.py 复制为 server1.py，并对其中的代码进行修改，修改后的代码如下：

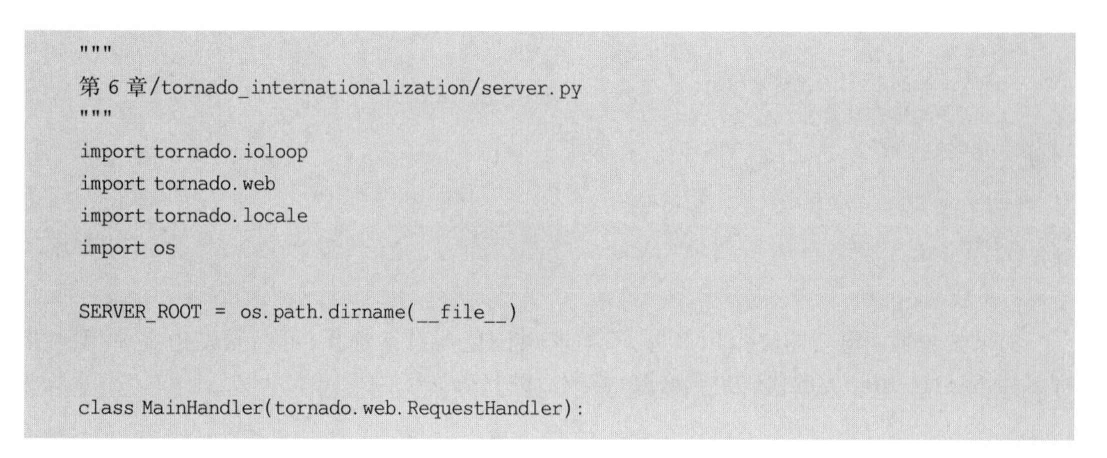

```
"""
第 6 章/tornado_internationalization/server.py
"""
import tornado.ioloop
import tornado.web
import tornado.locale
import os

SERVER_ROOT = os.path.dirname(__file__)


class MainHandler(tornado.web.RequestHandler):
```

```
    async def get(self):
        #渲染模板,并向其传入参数 greeting_word
        await self.render("main.html", greeting_word = "Hello")


if __name__ == "__main__":
    #从指定的目录中加载语言表
    tornado.locale.load_translations(os.path.join(SERVER_ROOT, "languages"))
    application = tornado.web.Application(
        [
            (r"/", MainHandler),
        ],
        #配置模板文件目录
        template_path = os.path.join(SERVER_ROOT, "template")
    )
    application.listen(8888)
    try:
        tornado.ioloop.IOLoop.current().start()
    except KeyboardInterrupt:
        print("User stopped server")
```

项目文件结构如图 6-24 所示。

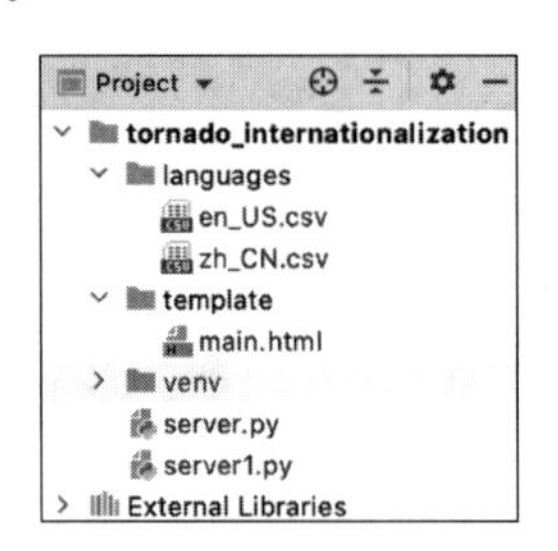

图 6-24 修改后的项目文件结构

在模板中使用_() 函数实现翻译功能，main.html 文件中的代码如下：

```
<!-- 第 6 章/tornado_internationalization/template/main.html -->
<!DOCTYPE html>
<html lang = "en">
<head>
<meta charset = "UTF - 8">
<title>模板渲染结果</title>
</head>
<body>
Greeting word: {{ _(greeting_word) }}
</body>
</html>
```

如果希望页面语言能够由用户选择，而非由浏览器自动选择，可以预先指定使用的语言，对 MainHandler 类的代码进行修改，修改后的代码如下：

```
class MainHandler(tornado.web.RequestHandler):
    async def get(self):
        # 根据 URL 中的参数 lang 来确定页面语言
        self.locale = tornado.locale.get(
            self.get_query_argument("lang", "en_US")
        )
        # 渲染模板,并向其传入参数 greeting_word
        await self.render("main.html", greeting_word = "Hello")
```

这段代码接收了参数 lang 作为页面的语言，如果不传参数，则默认使用英语，如果在代码中指定 lang=zh_CN，就算使用英文版的浏览器，也会呈现中文页面，效果如图 6-25 所示。

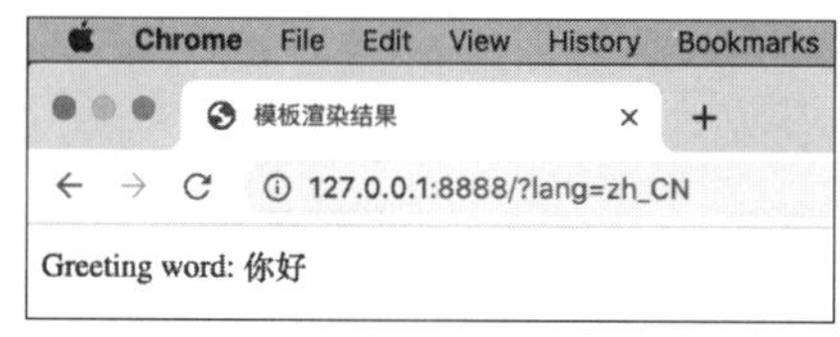

图 6-25 用户指定语言效果

6.7 使用 WSGIContainer 集成旧系统

虽然最近几年异步 IO 编程发展迅速，但目前 Python Web 的主流开发技术还是基于 WSGI，因此绝大多数网站是基于 WSGI 技术构建的。这类网站在向异步 IO 编程模型迁移的时候，如果立即全部更换技术显然是不现实的，成本极高，短期难以见到成效。这就要求新技术必须对旧技术提供支持，以便慢慢迁移，直到完全迁移完毕。

Tornado 框架提供了 WSGIContainer，用于支持旧系统，作为架构迁移过程中的临时方案极为重要。

下面以集成 web2py 为例讲解如何集成旧系统。

web2py 是一个极为强大的全栈开发框架，功能集成度极高，甚至可以让开发者在短短数天内开发出一个相对复杂的系统，将流行语“人生苦短，我用 Python”的理念发挥到了极致。

进入页面 http://web2py.com/init/default/download，下载源码版 web2py，操作如图 6-26 所示。

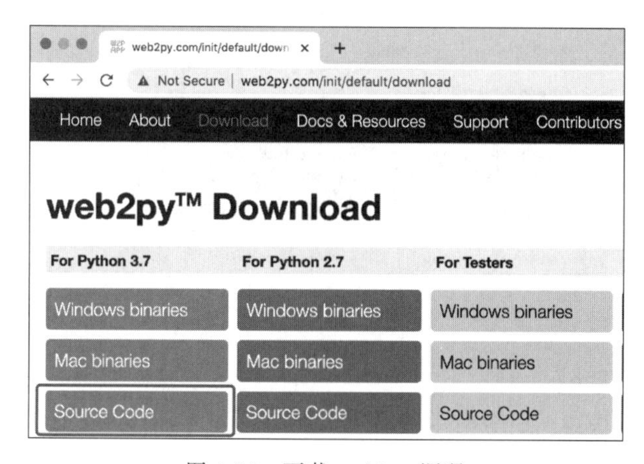

图 6-26 下载 web2py 源码

下载后解压并执行其中的 web2py. py 文件,如图 6-27 所示。

在 Windows 系统中,如果已经安装了 Python 环境,可以直接双击该文件进行启动。在 Linux 或者苹果系统中,则需要在命令行终端里通过命令 python web2py. py 执行,如图 6-28 所示。

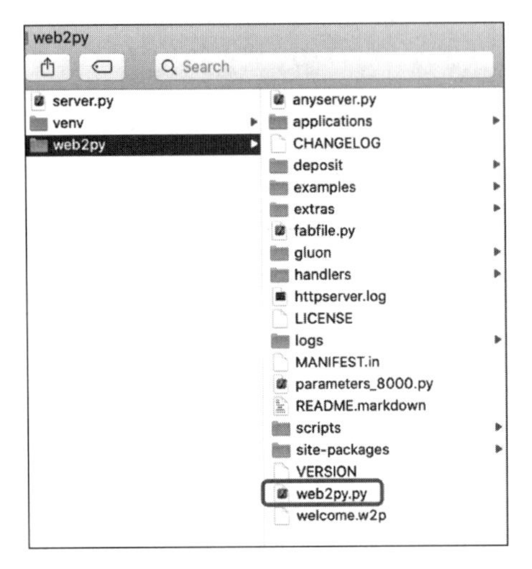

图 6-27 web2py 项目启动文件

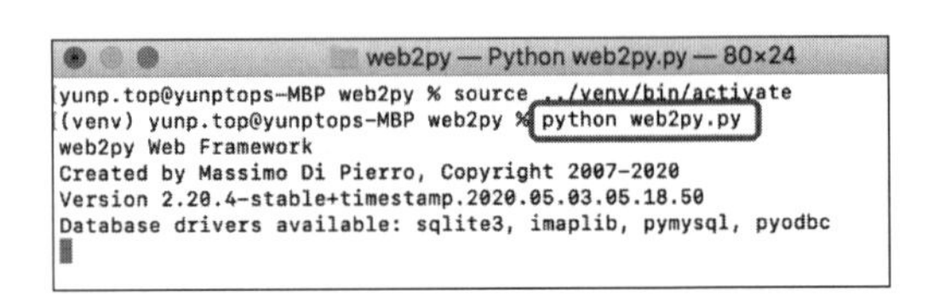

图 6-28 启动 web2py

该操作将启动 web2py 界面,如图 6-29 所示。

输入配置服务器管理面板密码,并单击 start server 按钮以启动服务器。也可以不使用该界面,而直接在终端里通过命令 python web2py -a pw -p 8000 -i 0. 0. 0. 0 启动服务器,如图 6-30 所示。

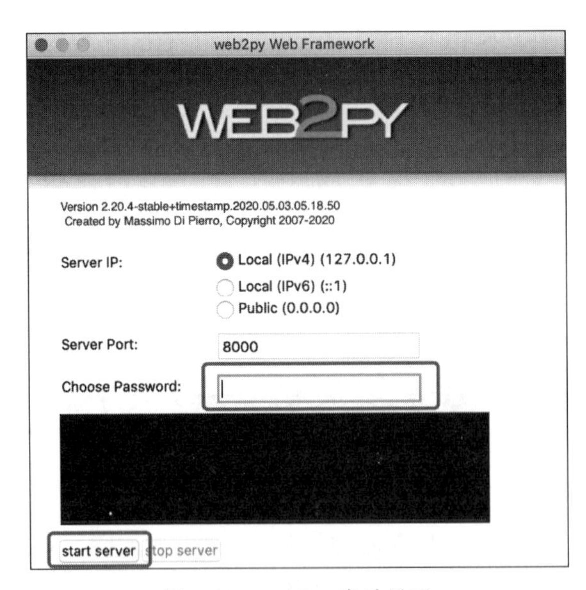

图 6-29 web2py 启动界面

该操作将在本机的 8000 端口启动 web2py 服务器,并且配置其管理面板密码为 pw,接下来在浏览器中通过 http://127. 0. 0. 1:8000 访问该服务器,如图 6-31 所示。

```
web2py — Python web2py.py -a pw -p 8000 -i 0.0.0.0 — 80×24
yunp.top@yunptops-MBP web2py % source ../venv/bin/activate
(venv) yunp.top@yunptops-MBP web2py % python web2py.py
web2py Web Framework
Created by Massimo Di Pierro, Copyright 2007-2020
Version 2.20.4-stable+timestamp.2020.05.03.05.18.50
Database drivers available: sqlite3, imaplib, pymysql, pyodbc
(venv) yunp.top@yunptops-MBP web2py % python web2py.py -a pw -p 8000 -i 0.0.0.0
web2py Web Framework
Created by Massimo Di Pierro, Copyright 2007-2020
Version 2.20.4-stable+timestamp.2020.05.03.05.18.50
Database drivers available: sqlite3, imaplib, pymysql, pyodbc

please visit:
        http://127.0.0.1:8000/
use "kill -SIGTERM 9905" to shutdown the web2py server
```

图 6-30 纯命令行启动 web2py 服务器

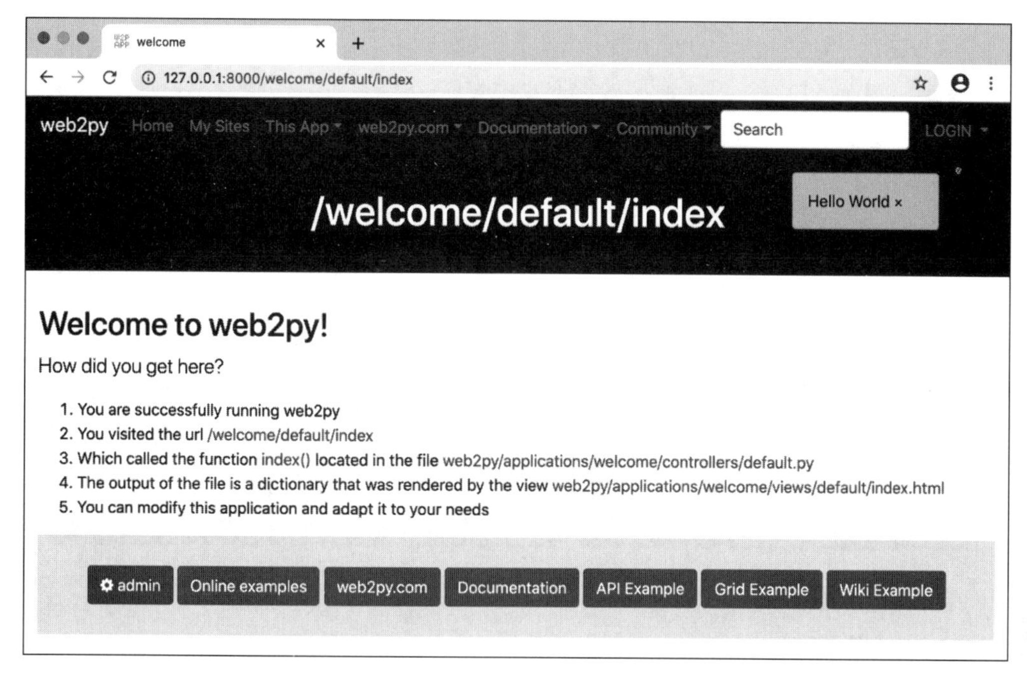

图 6-31 web2py 网站页面

接下来将 web2py 集成到 Tornado 中以讲解 WSGIContainer 的用法。项目文件结构如图 6-32 所示。

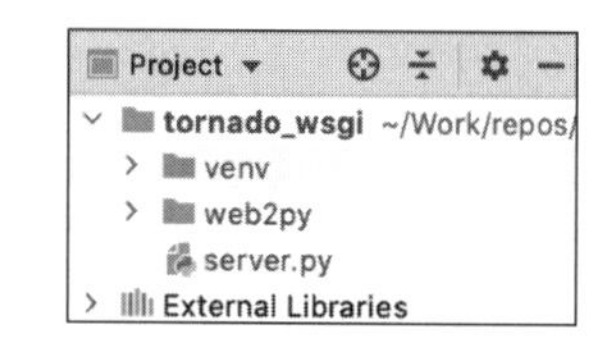

图 6-32 集成 WSGI 应用的项目文件结构

将 web2py/handlers/wsgihandler.py 文件复制到 web2py 目录，在 server.py 文件中输入代码如下：

```
"""
第 6 章/tornado_wsgi/server.py
"""

import tornado.wsgi
import tornado.httpserver
import tornado.ioloop
```

```
import web2py.wsgihandler

# 创建一个 WSGI 容器,用于支持 WSGI 应用
wsgi_container = tornado.wsgi.WSGIContainer(
    web2py.wsgihandler.application
)

if __name__ == '__main__':
    http_server = tornado.httpserver.HTTPServer(
        wsgi_container
    )
    http_server.listen(8888)
    try:
        tornado.ioloop.IOLoop.current().start()
    except KeyboardInterrupt:
        print("User stopped server")
```

启动该服务器后用浏览器通过 http://127.0.0.1:8888 进行访问，效果如图 6-33 所示。

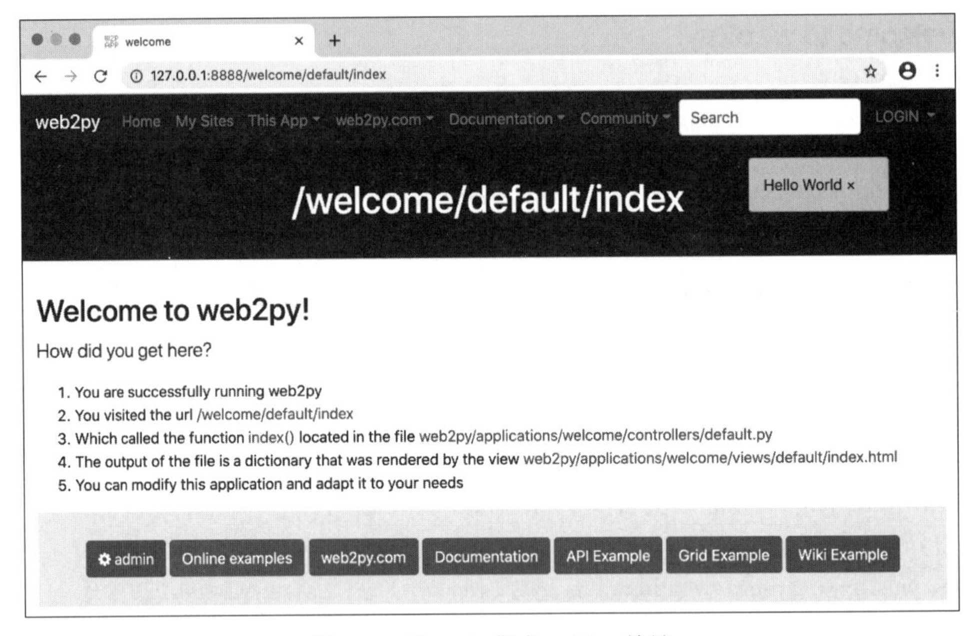

图 6-33　Tornado 集成 web2py 效果

如果要同时混合 Tornado 应用与 web2py 应用，可以使用 FallbackHandler 将特定的请求转发给 web2py 以便进行处理，对 server.py 代码进行修改，修改后的代码如下：

```
"""
第 6 章/tornado_wsgi/server1.py
"""

import tornado.ioloop
```

```
import tornado.web
import tornado.wsgi

import web2py.wsgihandler


class Home(tornado.web.RequestHandler):
    def get(self):
        self.write("Home page")


#创建一个 WSGI 容器,用于支持 WSGI 应用
wsgi_container = tornado.wsgi.WSGIContainer(
    web2py.wsgihandler.application
)

if __name__ == '__main__':
    http_server = tornado.web.Application([
        (r"/", Home),
        (
            r"/welcome.*",
            #使用 FallbackHandler 可将指定的请求交由 WSGI 应用处理
            tornado.web.FallbackHandler,
            dict(fallback = wsgi_container)
        )
    ])
    http_server.listen(8888)
    try:
        tornado.ioloop.IOLoop.current().start()
    except KeyboardInterrupt:
        print("User stopped server")
```

启动该服务器后,通过浏览器访问 http://127.0.0.1:8888 可以看到如图 6-34 所示页面。

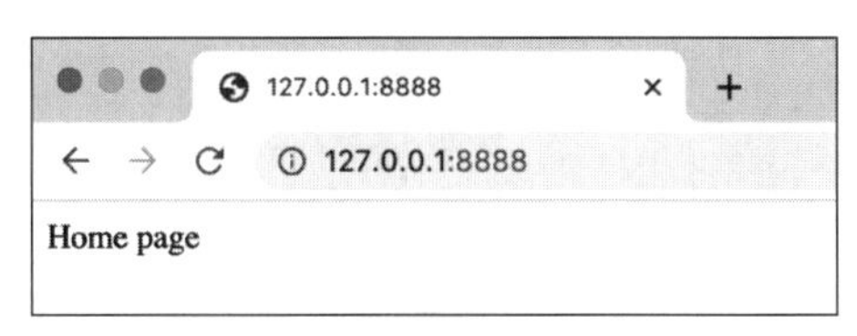

图 6-34 Tornado 处理请求

访问 http://127.0.0.1:8888/welcome 可以看到 web2py 的页面,如图 6-35 所示。

因为 Django 也支持 WSGI,所以 Tornado 也可以通过 WSGIContainer 集成 Django 应用,其原理相同,读者可自行尝试。

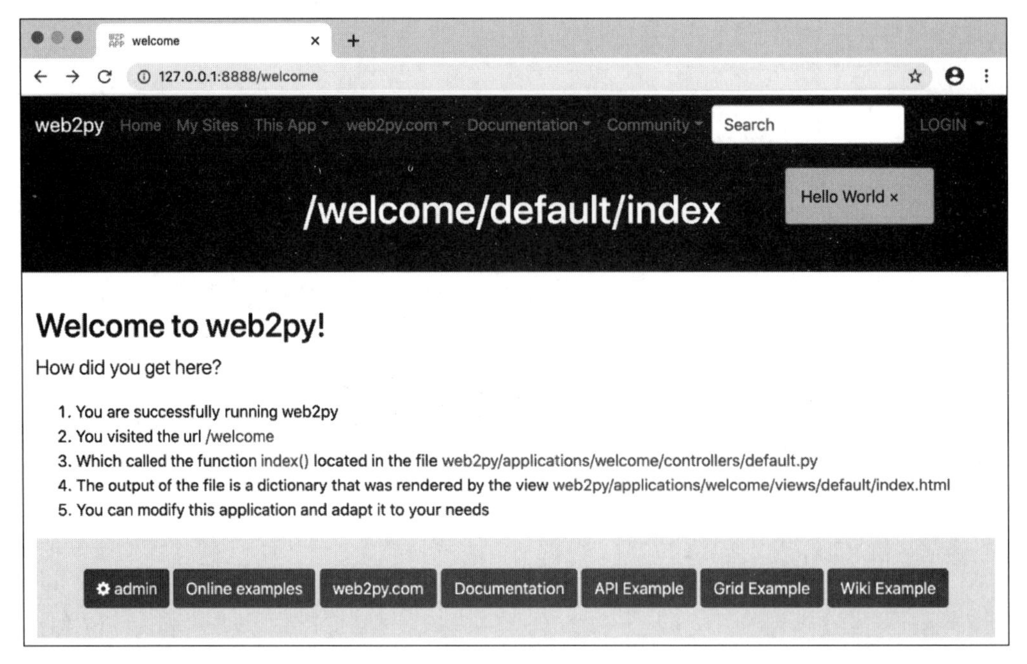

图 6-35 web2py 处理请求

6.8 HTTP 客户端

Tornado 提供了 HTTP 客户端，在实际开发工作中非常有用，使用起来也相当方便，首先准备服务器的源码，代码如下：

```
"""
第 6 章/tornado_httpclient/server.py
"""

import tornado.ioloop
import tornado.web


class MainHandler(tornado.web.RequestHandler):
    def get(self):
        # 获取以 GET 方式传来的参数
        name = self.get_query_argument("name", "")
        self.write(f"Hello {name}")

    def post(self):
        # 获取以 POST 方式传来的参数
        name = self.get_body_argument("name", "")
        self.write(f"Hello {name}")
```

```
if __name__ == "__main__":
    application = tornado.web.Application([
        (r"/", MainHandler),
    ])
    application.listen(8888)
    try:
        tornado.ioloop.IOLoop.current().start()
    except KeyboardInterrupt:
        print("User stopped server")
```

在该服务器中，实现了处理GET和POST请求的功能，接下来使用Tornado的HTTP客户端以GET方式请求服务器，代码如下：

```
"""
第6章/tornado_httpclient/client_get.py
"""

import tornado.httpclient
import asyncio

#创建一个基于异步IO的HTTP客户端
client = tornado.httpclient.AsyncHTTPClient()


async def main():
    #向服务器发送一个GET方式的请求
    resp = await client.fetch("http://127.0.0.1:8888?name=Xiaoming")
    #将服务器返回的结果以utf-8的编码方式进行解码
    print(resp.body.decode("utf-8"))


asyncio.run(main())
```

运行效果如图6-36所示。

```
server ×    client_get ×
/Users/yunp.top/Work/repos/github/
Hello Xiaoming

Process finished with exit code 0
```

图6-36 Tornado HTTP客户端发送GET方式的请求

也可使用Tornado HTTP客户端发送POST方式的请求，代码如下：

```
"""
第6章/tornado_httpclient/client_post.py
"""
import tornado.httpclient
```

```
import asyncio

# 创建一个基于异步 IO 的 HTTP 客户端
client = tornado.httpclient.AsyncHTTPClient()


async def main():
    # 向服务器发送请求
    resp = await client.fetch(
        # 创建一个 POST 方式的请求,并将参数传给服务器
        tornado.httpclient.HTTPRequest(
            "http://127.0.0.1:8888",
            "POST",
            # POST 方式的请求可以附加较大的数据(一般不超过 100MB)
            # 将参数写在 body 中发送给服务器
            body = "name = Xiaoming"
        )
    )
    print(resp.body.decode("utf - 8"))


asyncio.run(main())
```

第7章 Socket. IO

HTTP 最初建立在 Socket 短连接机制之上，这给实时通信类的功能实施带来了极大的困难。在 HTTP 1.1 标准中提供了长连接，目前市面上大多数服务器已经完全支持 HTTP 1.1 的标准了，同时浏览器也进入了 HTML 5 时代，提供了与 WebSocket 相关的 API，让基于浏览器的高效的、实时的通信成为现实。

虽然浏览器端对于 WebSocket 有良好的 API 支持，但是服务器端使用 WebSocket 还是有一定的难度，所以通常使用第三方框架 Socket. IO 实现实时通信类的功能。

Socket. IO 是一个实时通信框架，在支持 WebSocket 的浏览器上采用 WebSocket 作为底层的通信技术，在不支持 WebSocket 的浏览器上采用 HTTP 长连接作为底层通信技术。

7.1 WebSocket 实时通信

在学习 Socket. IO 之前应该先学习 WebSocket，虽然在实际开发工作中会优先选择 Socket. IO，并且确实会大大节省项目开发成本，但是其底层的 WebSocket 技术也有一定的优势。在项目中使用 WebSocket 技术更加直接、简洁、高效，并且在有些需求场景下有着不可替代的作用。

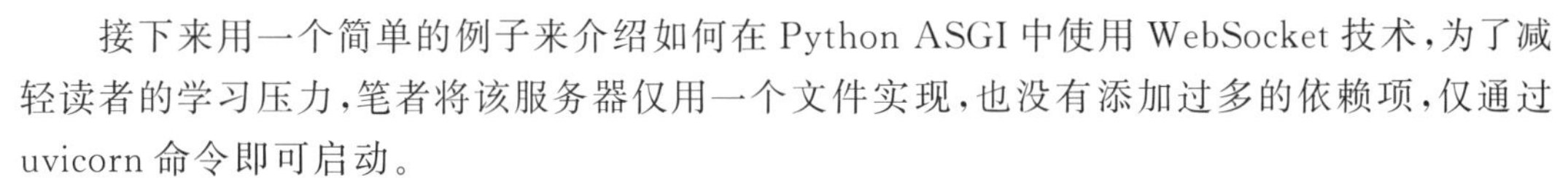

接下来用一个简单的例子来介绍如何在 Python ASGI 中使用 WebSocket 技术，为了减轻读者的学习压力，笔者将该服务器仅用一个文件实现，也没有添加过多的依赖项，仅通过 uvicorn 命令即可启动。

先看示例效果，如图 7-1 所示。

在页面中有一个按钮 Ping，单击该按钮向服务器发送一个消息，服务器实时回应消息 Pong。

项目文件结构如图 7-2 所示。

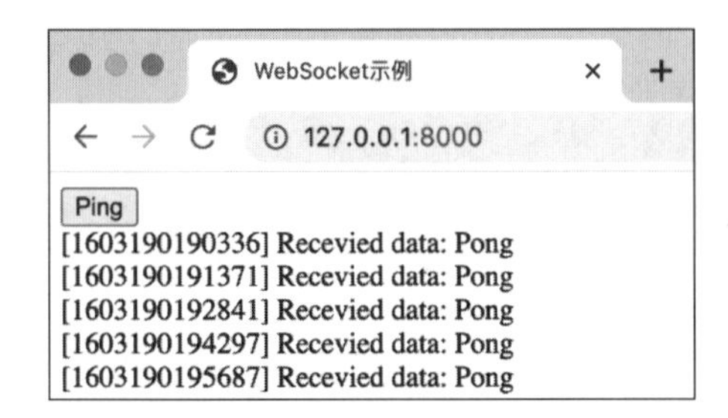

图 7-1　WebSocket 示例效果

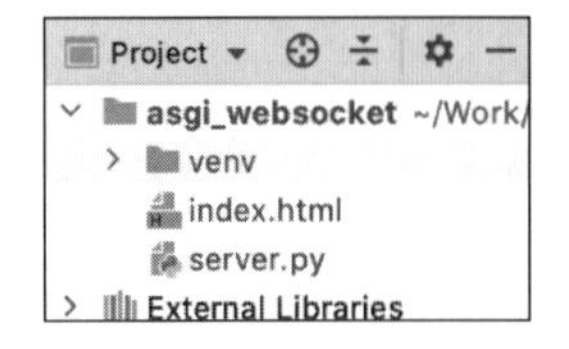

图 7-2　WebSocket 示例项目文件结构

当浏览器请求网站根目录时，服务器通过异步 IO 读取 index.html 文件并将其内容返回浏览器端，其中 server.py 文件代码如下：

```
"""
第 7 章/asgi_websocket/server.py
"""

import asyncio, os

SERVER_ROOT = os.path.dirname(__file__)


async def handle_lifespan(scope, receive, send):
    while True:
        # 不断读取数据
        message = await receive()
        # 如果读取消息类型为 lifespan.startup,则进行初始化操作
        if message['type'] == 'lifespan.startup':
            # 在初始化完成后,向 ASGI 环境发送启动完成消息
            await send({'type': 'lifespan.startup.complete'})
        # 如果读取消息类型为 lifespan.shutdown,则进行收尾工作
        elif message['type'] == 'lifespan.shutdown':
            # 在收尾工作结束后,向 ASGI 环境发送收尾完成消息
            await send({'type': 'lifespan.shutdown.complete'})
            break


async def handle_home_page_request(scope, receive, send):
    # 向浏览器发送 HTTP 协议头
    await send({
        'type': 'http.response.start',
        'status': 200,
        'headers': [
            [b'content-type', b'text/html'],
        ]
    })

    # 获取当前的事件循环
    loop = asyncio.get_running_loop()
    # 以异步 IO 的方式打开 index.html 文件,并配置打开模式
    # 为读取二进制数据
    f = await loop.run_in_executor(
        None, open,
        os.path.join(SERVER_ROOT, "index.html"), "rb"
    )
    # 以异步 IO 的方式读取文件
    data = await loop.run_in_executor(None, f.read)
    # 以异步 IO 的方式关闭文件
    await loop.run_in_executor(None, f.close)
```

```
    #向浏览器发送文件内容
    await send({
        'type': 'http.response.body',
        'body': data,
        'more_body': False
    })


async def handle_ws_conn(scope, receive, send):
    while True:
        msg = await receive()
        msg_type = msg['type']

        if msg_type == 'websocket.receive':
            #当接收到浏览器端的 Ping 消息时回应 Pong
            if msg['text'] == 'Ping':
                await send({'type': 'websocket.send', 'text': "Pong"})

        #该消息类型表示 WebSocket 建立
        elif msg_type == 'websocket.connect':
            print("Client connected")
            #向浏览器发送 websocket.accept 消息,表示接受该连接
            await send({'type': 'websocket.accept'})

        #该消息类型表示失去了 WebSocket 连接
        if msg_type == 'websocket.disconnect':
            print("Client disconnected")
            break


async def send_404_error(scope, receive, send):
    await send({
        'type': 'http.response.start',
        'status': 404,
        'headers': [
            [b'content-type', b'text/html'],
        ]
    })
    await send({
        'type': 'http.response.body',
        'body': b"Not found",
        'more_body': False
    })


async def app(scope, receive, send):
    #获取请求类型
    request_type = scope['type']

    #如果是 http 类型的请求,则由该程序段处理
```

```
    if request_type == 'http':
        #获取请求的路径
        request_path = scope['path']
        if request_path == "/":
            await handle_home_page_request(scope, receive, send)
        else:
            #对于其他路径的请求,均向浏览器发回 404 错误
            await send_404_error(scope, receive, send)

    #如果是 websocket 类型的请求,则由该程序段处理
    elif request_type == 'websocket':
        await handle_ws_conn(scope, receive, send)

    #如果是生命周期类型的请求,则由该程序段处理
    elif request_type == 'lifespan':
        await handle_lifespan(scope, receive, send)
    else:
        raise NotImplementedError()
```

index.html 文件的代码如下：

```
<!-- 第 7 章/asgi_websocket/index.html -->
<!DOCTYPE html>
<html lang="en">

<head>
<meta charset="UTF-8">
<title>WebSocket 示例</title>
</head>

<body>

<div>
<button>Ping</button>
</div>
<div id="output">
</div>

<script>
        (function () {
            let output = document.querySelector("#output");

            //创建 WebSocket 实例,用于连接服务器
            let ws = new WebSocket(`ws://${location.host}`);

            //当收到服务器消息时触发该事件
            ws.onmessage = e => {
                output.innerHTML += `[${Date.now()}] Recevied data: ${e.data}<br>`;
            };

            //为按钮添加单击侦听器,在单击 Ping 按钮时向服务器发送 Ping
            document.querySelector("button").onclick = e => {
```

```
                    ws.send("Ping");
                };
            })();
    </script>
</body>

</html>
```

若要运行该项目，需要先使用命令 pip install uvicorn websockets 安装依赖项，然后使用命令 uvicorn server:app 启动服务器。

7.2 Socket.IOASGIApp

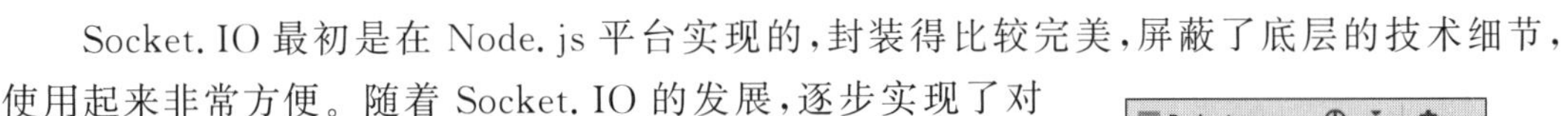

Socket.IO 最初是在 Node.js 平台实现的，封装得比较完美，屏蔽了底层的技术细节，使用起来非常方便。随着 Socket.IO 的发展，逐步实现了对其他编程语言的支持，目前在 Python 语言中也有一个 Socket.IO 的实现，叫作 python-socketio。如果要在 Python 语言中使用 Socket.IO 框架，需要先通过命令 pip install python-socketio 安装该依赖项。

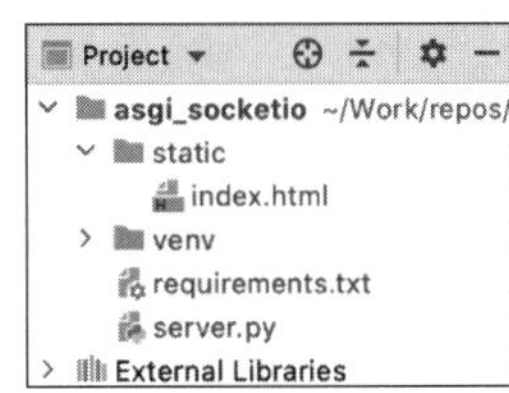

图 7-3 ASGIApp 项目文件结构

项目文件结构如图 7-3 所示。

其中 server.py 文件的代码如下：

```
"""
第 7 章/asgi_socketio/server.py
"""

import socketio, os

SERVER_ROOT = os.path.dirname(__file__)

#创建一个 AsyncServer 实例
sio = socketio.AsyncServer()

app = socketio.ASGIApp(sio, static_files = {
    #配置静态文件目录为服务器应用目录下的 static 目录,这
    #样可以使用 Socket.IO 内置的静态文件服务器处理静态请求
    "/static": os.path.join(SERVER_ROOT, "static")
})
```

index.html 文件的代码如下：

```
<!-- 第 7 章/asgi_socketio/static/index.html -->
<!DOCTYPE html>
<html lang="en">
<head>
```

```
<meta charset="UTF-8">
<title>Title</title>
</head>
<body>
这是一个静态文件
</body>
</html>
```

使用命令 uvicorn server:app 启动服务器，如图 7-4 所示。

```
Terminal:   Local ×   +
(venv) yunp.top@yunptops-MBP asgi_socketio % uvicorn server:app
INFO:     Started server process [25052]
INFO:     Waiting for application startup.
INFO:     Application startup complete.
INFO:     Uvicorn running on http://127.0.0.1:8000 (Press CTRL+C to quit)
```

图 7-4 启动服务器

在浏览器中通过地址 http://127.0.0.1:8000/static/index.html 进行访问，效果如图 7-5 所示。

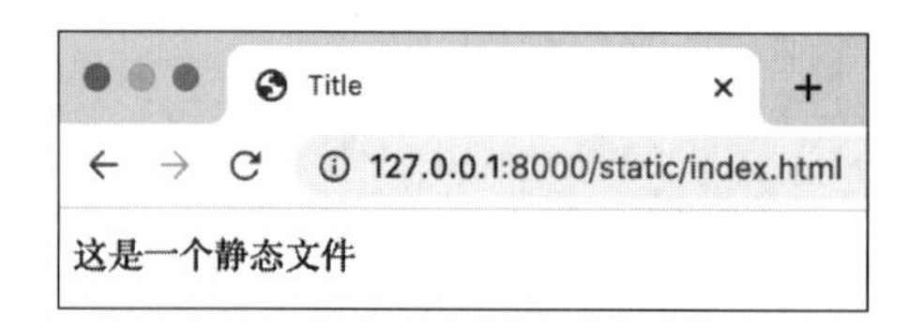

图 7-5 访问静态页面显示效果

7.3 Socket.IO 实时通信

在 Socket.IO 中如果需要侦听浏览器连接事件，只需为函数配上 event 装饰器。例如要侦听 my_event 事件，则函数代码如下：

```
@sio.event
async def my_event(sid, data):
    pass
```

也可以使用 on 装饰器实现，代码如下：

```
@sio.on("my_event")
async def my_event(sid):
    pass
```

浏览器端若要触发该事件，则可通过如下代码实现：

```
socket.emit("my_event");
```

服务器端可向该 sid 发送回应消息，代码如下：

```
await sio.emit("my_event_callback", to = sid)
```

如果只有一次通信需要回应，则可以使用回调机制，例如浏览器端发起了一个登录请求，代码如下：

```
//向服务器发送登录事件，并将回调函数作为最后一个参数
socket.emit("login", { name: "yunp" }, result => {
});
```

服务器端处理该请求的代码如下：

```
@sio.event
async def login(sid, data):
    #该函数的返回值作为浏览器端回调函数的传入参数传给浏览器端
    return f"Hello {data['name']}, you are granted!"
```

接下来实现一个完整的前后端通信项目，项目文件结构如图 7-6 所示。

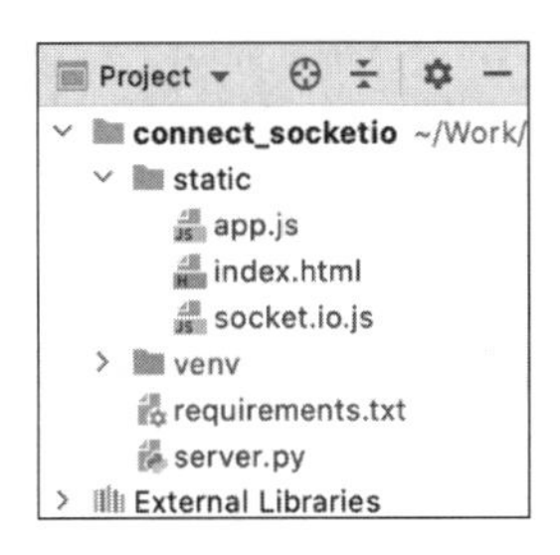

图 7-6　Socket.IO 通信项目文件结构

其中 server.py 的代码如下：

```
"""
第 7 章/connect_socketio/server.py
"""

import socketio, os

SERVER_ROOT = os.path.dirname(__file__)

#创建一个 AsyncServer 实例
sio = socketio.AsyncServer(
    #配置该服务器的异步模式为 asgi
    async_mode = 'asgi',
    #配置该服务器，使其允许来自所有域的连接
    cors_allowed_origins = "*"
```

```
)


@sio.event
async def my_event(sid):
    pass
    """
侦听浏览器端发来的 Ping 消息,并向该连接发回 Pong 消息
    :param sid: Socket.IO 为该连接编码的唯一标识
    :param data: 通过该事件传来的数据
    :return:
    """

    #向该连接发回 Pong 消息
    await sio.emit("my_event_callback", to=sid)


@sio.event
async def login(sid, data):
    #该函数的返回值作为浏览器端回调函数的传入参数传给浏览器端
    return f"Hello {data['name']}, you are granted!"


app = socketio.ASGIApp(sio, static_files={
    #配置静态文件目录为服务器应用目录下的 static 目录,这
    #样可以使用 Socket.IO 内置的静态文件服务器处理静态请求
    "/static": os.path.join(SERVER_ROOT, "static")
})
```

依赖项列表 requirements.txt 内容如下:

```
python-socketio==4.6.0
uvicorn==0.12.2
websockets==8.1
```

index.html 文件代码如下:

```
<!-- 第7章/connect_socketio/static/index.html -->
<!DOCTYPE html>
<html lang="en">

<head>
<meta charset="UTF-8">
<title>Title</title>
<script src="socket.io.js"></script>
</head>

<body>
```

```
<div>
<button class="btn_my_event">my_event</button>
<button class="btn_login">Login</button>
</div>
<div id="output">
</div>

<script src="app.js"></script>
</body>

</html>
```

app.js 文件代码如下：

```
//第 7 章/connect_socketio/static/app.js

(function () {
    let socket = io();

    let output = document.querySelector("#output");

    //侦听服务器端的 my_event_callback 事件
    socket.on("my_event_callback", e => {
        output.innerHTML += `[${Date.now()}] my_event_callback from server<br>`;
    });

    document.querySelector(".btn_my_event").onclick = e => {
        //向服务器派发 my_event 事件
        socket.emit("my_event");
    };

    document.querySelector(".btn_login").onclick = e => {
        //向服务器发送登录事件,并将回调函数作为最后一个参数
        socket.emit("login", { name: "yunp" }, result => {
            output.innerHTML += `[${Date.now()}] ${result}<br>`;
        });
    };
})();
```

socket.io.js 文件可从 Socket.IO 官网 https://socket.io/或者官方提供的 cdn 地址 https://cdnjs.com/libraries/socket.io 进行下载。

启动服务器后在浏览器中通过 http://127.0.0.1:8000/static/index.html 访问页面，分别单击 my_event 和 Login 按钮，可以看到效果如图 7-7 所示。

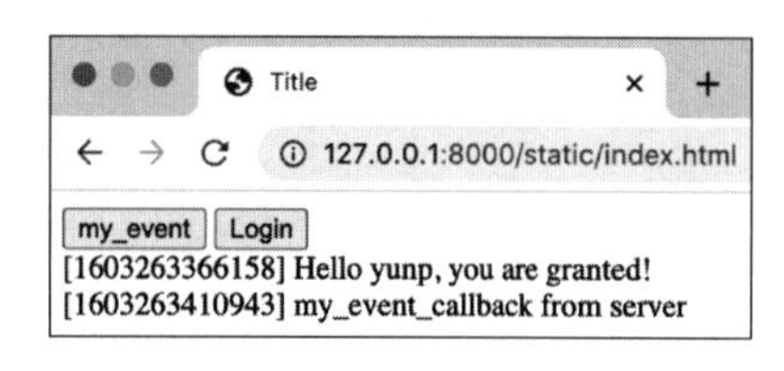

图 7-7　Socket.IO 通信效果

7.4 实现聊天室服务器端

服务器端可使用 sio.emit 函数转发浏览器端的消息，使用 to 参数可向指定的浏览器端或者房间发送消息，如果将 to 参数置空，则向所有连接的浏览器端发送消息。在该示例中，我们以最简单的方式来学习 Socket.IO 多终端实时通信的知识，核心代码如下：

```
#侦听浏览器端发来的 msg 事件
@sio.event
async def msg(sid, content):
    #向所有已连接的终端广播消息
    await sio.emit('msg', {'from': sid, 'content': content})
```

已连接的浏览器端可向该服务器发送 msg 消息，服务器直接向所有终端广播该消息，并附加发送消息的终端 id。

完整的代码如下：

```
"""
第 7 章/chatroom/server.py
"""

import socketio, os

SERVER_ROOT = os.path.dirname(__file__)

#创建一个 AsyncServer 实例
sio = socketio.AsyncServer(
    #配置该服务器的异步模式为 asgi
    async_mode = 'asgi'
)


#侦听浏览器端发来的 msg 事件
@sio.event
async def msg(sid, content):
    #向所有连接终端广播消息
    await sio.emit('msg', {'from': sid, 'content': content})


#创建一个 ASGIApp 应用,可使用 uvicorn 等协议层实现调用
app = socketio.ASGIApp(
    sio,
    static_files = {
        #配置静态文件目录为服务器应用目录下的 static 目录,这
        #样可以使用 Socket.IO 内置的静态文件服务器处理静态请求
        "/static": os.path.join(SERVER_ROOT, "static")
    }
)
```

7.5 实现聊天室浏览器端

与服务器端的实现相比，浏览器端的实现相对复杂一些，毕竟需要开发界面，但整体来讲并不难，相比直接使用 WebSocket API 实现省去了大量的工作。

使用 Socket.IO 框架，可以省去的主要工作如下：

(1) 不用关心连接地址，因为已经有了默认值作为当前的服务器地址。

(2) 如果要支持复杂的系统，还可以连接不同的路径(socketio_path)，每个路径可以理解为一个单独的应用，而且在同一个应用中还可以用命名空间进行分组。

(3) 不用考虑如何标识连接终端，因为 Socket.IO 会自动为每个连接终端生成唯一编码。

(4) 不用设计自定义的协议用于区分消息的用途，因为一切消息都被设计为事件通知的形式，不同用途的消息只不过是不同的事件，结构清晰、用法合理，代码可读性高、易于维护。

(5) 不用设计房间机制，因为已经实现了。

(6) 不用设计远程函数调用的回调机制，因为已经实现了，而且用法很简单。

项目文件结构如图 7-8 所示。

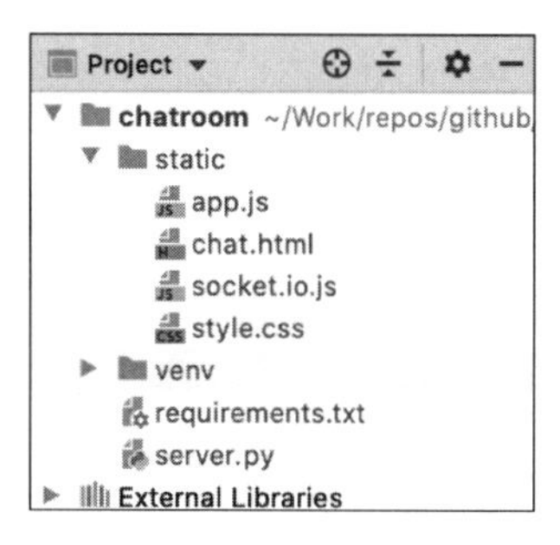

图 7-8 聊天室项目文件结构

其中 chat.html 文件代码如下：

```
<!-- 第 7 章/chatroom/static/chat.html -->
<!DOCTYPE html>
<html lang="en">

<head>
<meta charset="UTF-8">
<title>简易聊天室</title>

<script src="socket.io.js"></script>
<link rel="stylesheet" href="style.css">
</head>

<body>
<div style="width: 400px;">
<div id="output">
</div>
<div class="input-control">
<input type="text" id="input">
<button id="btn-send">Send</button>
</div>
</div>

<script src="app.js"></script>
```

```
</body>

</html>
```

文件 style.css 代码如下：

```
/* 第 7 章/chatroom/static/style.css */

#output {
    width: 100%;
    height: 300px;
    border: solid;
    border-width: thin;
    overflow: auto;
}

.input-control {
    width: 100%;
    margin-top: 5px;
    display: flex;
    flex-direction: row;
}

.input-control input {
    flex: 1;
    margin-right: 5px;
}
```

文件 app.js 代码如下：

```
//第 7 章/chatroom/static/app.js

(function () {

    let input, output, btnSend;
    let socket;

    function findElements() {
        input = document.querySelector("#input");
        output = document.querySelector("#output");
        btnSend = document.querySelector("#btn-send");
    }

    function tryToSendMsg() {
        let msg = input.value;
        if (msg) {
            //向服务器发送消息
            socket.emit("msg", msg);
            //清空输入框
            input.value = "";
```

```
        }
    }

    function btnSend_clickedHandler() {
        tryToSendMsg();
    }

    function input_keyupHandler(e) {
        //当按下回车键时发送数据
        if (e.key == "Enter") {
            tryToSendMsg();
        }
    }

    /**
     * 处理 msg 事件
     * @param {*} msg 服务器发来的消息
     */
    function socket_msgHandler(msg) {
        output.innerHTML += `${msg.from}: ${msg.content}<br>`;

        //将文字滚动到底端
        output.scrollTop = output.scrollHeight;
    }

    function addListeners() {
        btnSend.onclick = btnSend_clickedHandler;
        input.onkeyup = input_keyupHandler;

        //侦听服务器发来的 msg 事件
        socket.on("msg", socket_msgHandler);
    }

    function connectServer() {
        /*
连接服务器，Socket.IO 默认连接的路径是 /socket.io,可通过
io({path:"/rt"}) 这种方式修改要连接的路径,则对应的服务器
端也应该将服务器启动在 /rt 路径上,对应的 Python ASGI 应用
代码类似 app = socketio.ASGIApp(sio,socketio_path="/rt")
         */
        socket = io();
    }

    function main() {
        findElements();
        connectServer();
        addListeners();
    };

    main();
})();
```

服务器启动成功后，可通过 http://127.0.0.1:8000/static/chat.html 地址进行访问，支持同时开启多个浏览器页面访问该地址，并且可以互相收发消息，如图 7-9 所示。

图 7-9　多终端聊天效果

7.6　Socket.IO 与 AIOHTTP 集成

如果 Socket.IO 要与 AIOHTTP 集成，需要将异步模式(async_mode)设置为 aiohttp，代码如下：

```
sio = socketio.AsyncServer(async_mode = 'aiohttp')
```

然后通过 attach 函数与 AIOHTTP 服务器关联，代码如下：

```
sio.attach(app)
```

该项目文件结构如图 7-10 所示。

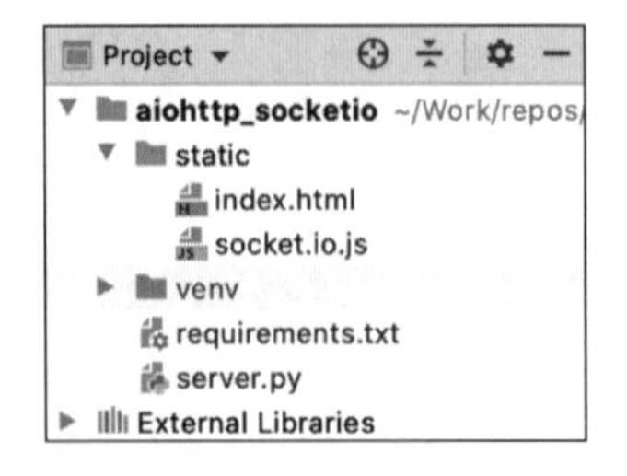

图 7-10　集成 AIOHTTP 项目文件结构

文件 server.py 的代码如下：

```
"""
第 7 章/aiohttp_socketio/server.py
"""

from aiohttp import web
import os, socketio, asyncio, time

SERVER_ROOT = os.path.dirname(__file__)

#配置 AIOHTTP >>>>>>>>>>>>>>>>>>>>>>>>>>>>>>>>>>>
#声明 aiohttp 的路由表
routes = web.RouteTableDef()


#将对网站根路径的访问用该函数处理
@routes.get("/")
async def home_page(request):
    #将页面重定向到 /static/index.html 页面
    raise web.HTTPTemporaryRedirect("/static/index.html")


#<<<<<<<<<<<<<<<<<<<<<<<<<<<<<<<<<<<<<<<<<<<<<<<<<<<<

#配置 socketio >>>>>>>>>>>>>>>>>>>>>>>>>>>>>>>>>>>
#创建 socketio 服务器,异步模式配置为 aiohttp
sio = socketio.AsyncServer(async_mode = 'aiohttp')


async def echo_task(sid):
    for i in range(1, 6):
        #向浏览器端发送消息
        await sio.emit(
            "echo",
            f"[{time.strftime('%X')}] Count: {i}",
            sid
        )
        #等待 1s
        await asyncio.sleep(1)


@sio.event
async def connect(sid, environ):
    """
连接成功
    """
    asyncio.create_task(echo_task(sid))
```

```
#<<<<<<<<<<<<<<<<<<<<<<<<<<<<<<<<<<<<<<<<<<<<<<<<<<<<<<<<<<<<

if __name__ == '__main__':
    #创建 aiohttp 服务器应用
    app = web.Application()

    #将 aiohttp 服务器与 socketio 服务器关联
    sio.attach(app)

    #添加静态文件目录
    routes.static("/static", os.path.join(SERVER_ROOT, "static"))
    app.add_routes(routes)
    web.run_app(app)
```

文件 index.html 的代码如下：

```
<!-- 第 7 章/aiohttp_socketio/static/index.html -->
<!DOCTYPE html>
<html lang="en">

<head>
<meta charset="UTF-8">
<title>Title</title>

<script src="socket.io.js"></script>
</head>

<body>
<script>
        (function () {
            let socket = io();
            socket.on("echo", d => console.log(d));
        })();
</script>
</body>

</html>
```

服务器会在浏览器连接成功之后向浏览器发送 5 条数据，每条数据间隔为 1s，效果如图 7-11 所示。

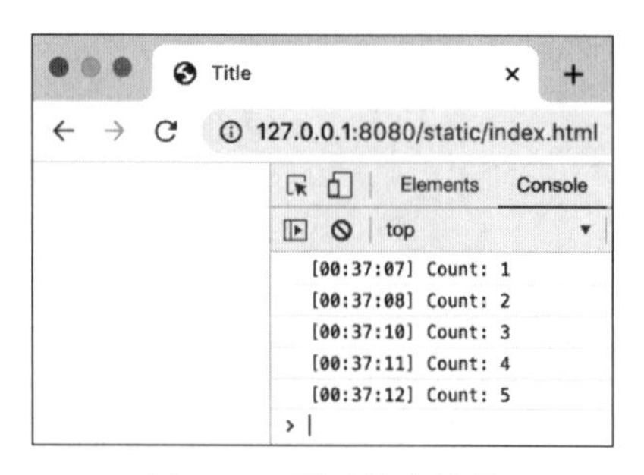

图 7-11 页面输出效果

7.7 Socket.IO 与 Tornado 集成

Tornado 与 AIOHTTP 是同类型的技术框架，都有稳定的异步开发支持 API。Socket.IO 是后起之秀，必然对所有异步 IO 技术框架都有着良好的支持，Tornado 也不例外。

如果要将 Tornado 与 Socket.IO 集成，在创建 Socket.IO 服务器时需要指定 async_mode 为 tornado，代码如下：

```
sio = socketio.AsyncServer(async_mode='tornado')
```

同时需要将该服务器映射到 Tornado 服务器的/socket.io/路由上，代码如下：

```
(r"/socket.io/", socketio.get_tornado_handler(sio))
```

完整的示例代码如下：

```
"""
第 7 章/aiohttp_socketio/server.py
"""

import os, socketio, asyncio, time
import tornado.web
import tornado.ioloop

SERVER_ROOT = os.path.dirname(__file__)

#创建 socketio 服务器,异步模式配置为 tornado
sio = socketio.AsyncServer(async_mode='tornado')


async def echo_task(sid):
    for i in range(1, 6):
        #向浏览器端发送消息
        await sio.emit(
            "echo",
            f"[{time.strftime('%X')}] Count: {i}",
            sid
        )
        #等待 1s
        await asyncio.sleep(1)


@sio.event
async def connect(sid, environ):
    """
连接成功
    """
```

```
    asyncio.create_task(echo_task(sid))


if __name__ == '__main__':
    #创建 tornado 服务器
    app = tornado.web.Application([
        #将 socketio 应用映射到 /socket.io/ 路径
        (r"/socket.io/", socketio.get_tornado_handler(sio)),
        (
            r"/static/(.*)",
            tornado.web.StaticFileHandler,
            {"path": os.path.join(SERVER_ROOT, "static")}
        )
    ])

    try:
        app.listen(8000)
        tornado.ioloop.IOLoop.current().start()
    except KeyboardInterrupt:
        print("User stopped server")
```

实 战 篇

实战篇包括第8章和第9章，其中第8章以ASGI技术为基础实现一个完整的异步IO全栈框架，第9章用一个实际的项目来演示如何使用异步IO技术开发一个完整的网站。

第8章 实现全栈框架cms4py

在基础篇里所讲的所有知识都是为亲手实现一个全栈框架做铺垫。异步编程已经出现了很多年，但是ASGI技术才刚刚开始起步。新生事物的发展力量总是非常强大，得益于异步编程技术多年的发展经验，ASGI从一开始就可以避免已知的问题，从而设计得非常完美。

相比Tornado与AIOHTTP来讲，ASGI分层明确，协议层可以自由更换。在项目进行技术迁移时有着不可替代的优势，尤其在Python异步编程的起步与发展阶段，各种技术的发展前景并不明朗，免不了后期有更新协议层的需求。

而应用层技术的发展正处于百花齐放的状态，虽然Quart和Daphne很优秀，但未来究竟会怎么样，还说不好。另外虽然Django也支持了ASGI，但是由于Django生态极为庞大，显得尾大不掉，短期内难以有革命性的改变，不过Django官方对ASGI技术热情极高，还主动推动着ASGI技术的发展，所以Django未来可期。

在目前这个窗口期，开发一款自己的应用层框架显得很有意义。首先不用担心后期更换应用层，其次可以积累技术及锻炼能力以提高自身的核心竞争力。笔者已经构建了一款基于ASGI技术的应用层框架，命名为cms4py。

由于cms4py的源码较为庞大，本章将根据核心原理及最初版本实现展开讲解。框架现已开源，地址为https://GitHub.com/cms4py/cms4py。在此笔者呼吁有志青年一起参与到这个开源项目中，在提高自己编程能力的同时携手为人类的开源事业作出自己的贡献，提升祖国的国际影响力。

8.1 制订需求

27min

在开发之前，需要制订一个简单的计划，为了简化实现以便高效讲解原理的运用，这里将制订最核心的功能需求。

（1）实现静态文件请求处理功能。静态文件请求处理是一个服务器应有的基本功能，也许在最终部署时会采用Apache或者Nginx等来处理静态文件请求，但服务器内置一个静态文件请求处理功能在开发阶段将起到非常重要的作用。

（2）动态页面请求处理功能。由于一般来讲动态页面需要与数据库进行交互，而网站的某个动态页面的内容常常很久都不会发生变化，这意味着如果每一次请求都要与数据库进行交互，将是巨大的资源浪费，所以实现动态页面请求处理功能的同时应该实现缓存机制，从而减少与数据库交互而产生的资源浪费。除了便于小功能改进而进行轻量级更新部署，还应该实现动态页面热更新功能。

（3）模板渲染功能。模板渲染功能可用于快速开发网站页面，是非常重要的功能，但是由于模板渲染涉及语法分析等技术，实现难度较大，幸运的是，可以通过集成现有的成熟模板引擎（如 Jinja2）快速实现。

（4）支持多语言功能。

（5）基于异步 IO 的数据库连接管理功能。与数据库操作相关的功能不难实现，在 Python 平台已经存在了很多 ORM/DAL 框架可供使用，但由于这些成熟框架都是基于阻塞型 IO 模型实现的，并不适合在 ASGI 中应用，所以实现该功能需要对现有的框架进行改良。

（6）实现 Session 机制。Session 是 Web 应用中非常重要的机制，可以在服务器端保存用户信息，在支持用户登录类的功能时不可或缺。

（7）集成 Socket.IO 以实现支持实时通信功能。

8.2 接入 ASGI

创建一个项目，名为 cms4py_first_generation，此时项目文件结构如图 8-1 所示。

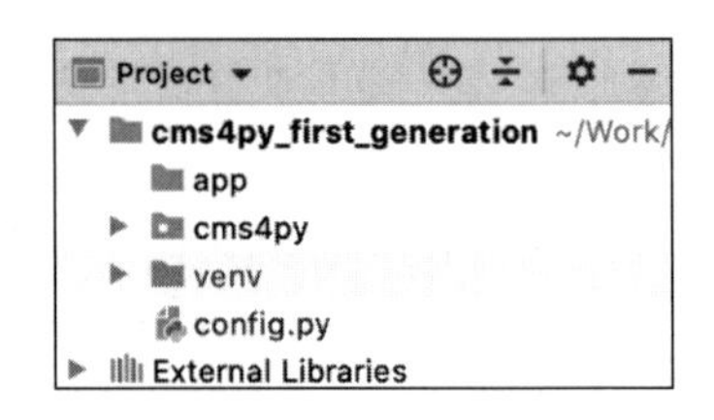

图 8-1 cms4py 初始项目文件结构

其中 cms4py 目录用于存放框架核心程序文件，app 目录用于存放用户应用程序文件，config.py 文件用于配置项目的重要参数，其代码如下：

```
"""
第 8 章/cms4py_first_generation/config.py
"""

import os

#服务器版本
SERVER_VERSION = '2020.02.15'
#服务器根目录
SERVER_ROOT = os.path.dirname(os.path.abspath(__file__))
#应用所在目录的目录名
APP_DIR_NAME = "app"
#应用根目录
APP_ROOT = os.path.join(SERVER_ROOT, APP_DIR_NAME)
#控制器所在目录的目录名
CONTROLLERS_DIR_NAME = "controllers"
#控制器所在目录
CONTROLLERS_ROOT = os.path.join(APP_ROOT, CONTROLLERS_DIR_NAME)
#静态文件根目录
STATIC_FILES_ROOT = os.path.join(APP_ROOT, "static")
#模板文件根目录
```

```
VIEWS_ROOT = os.path.join(APP_ROOT, 'views')
#语言文件根目录
LANGUAGES_ROOT = os.path.join(APP_ROOT, 'languages')
#默认语言
LANGUAGE = None #"zh-CN", "en-US"

#服务器名称
SERVER_NAME = 'cms4py'

#应用名称
APP_NAME = "cms4py"

"""
日志级别

CRITICAL    50
ERROR       40
WARNING     30
INFO        20
DEBug       10
NOTSET      0
"""
LOG_LEVEL = 10

#在用户未指定控制器的情况下由该默认控制器接收请求
DEFAULT_CONTROLLER = "default"
#在用户未指定控制器函数的情况下由该动作(action)接收请求
DEFAULT_ACTION = 'index'

#应用版本
APP_VERSION = '2020.02.15'

#全局编码方式
GLOBAL_CHARSET = 'utf-8'

#在 Cookie 中存储 Session ID 所使用的键名
CMS4PY_SESSION_ID_KEY = b"cms4py_session_id"
```

可暂时不用过于关心每个参数的用途,在下面的章节中会一一进行讲解。

接下来实现日志系统。在大型项目中,良好的日志工具与格式能够大大降低排查错误的难度,所以日志系统极为重要,每一条日志均应包括相关的所有关键信息。日志工具源码如下:

```
"""
第 8 章/cms4py_first_generation/cms4py/utils/log.py
"""

import logging
import config
```

```
class Cms4pyLog:
    __instance = None

    @staticmethod
    def get_instance():
        if not Cms4pyLog.__instance:
            Cms4pyLog.__instance = Cms4pyLog()
        return Cms4pyLog.__instance

    def __init__(self) -> None:
        super().__init__()
        self._log = logging.getLogger(config.APP_NAME)

        log_handler = logging.StreamHandler()
        log_handler.setFormatter(
            #设置自定义的日志格式,输出的日志包括行号等关键信息
            logging.Formatter(
                "[%(levelname)s %(name)s %(asctime)s %(pathname)s(%(lineno)s)] %
(message)s"
            )
        )

        if self._log.parent:
            self._log.parent.handlers = []
            #设置自定义的日志处理工具
            self._log.parent.addHandler(log_handler)

        #设置日志的级别,该参数来自 config.py 文件中的配置,用于过滤不
        #希望出现的日志
        self._log.setLevel(config.LOG_LEVEL)
        self.info = self._log.info
        self.debug = self._log.debug
        self.warning = self._log.warning
        self.error = self._log.error
```

为了简化第一个程序的业务逻辑,现对所有的请求均返回一个404页面,主程序源码如下:

```
"""
第8章/cms4py_first_generation/cms4py/__init__.py
"""

from cms4py.handlers import lifespan_handler
from cms4py.handlers import error_pages
from cms4py.utils.log import Cms4pyLog


async def application(scope, receive, send):
    #获取请求类型
```

```
    request_type = scope['type']

    # 如果是 http 类型的请求,则由该程序段处理
    if request_type == 'http':
        # 对于未被处理的请求,均向浏览器发回 404 错误
        await error_pages.send_404_error(scope, receive, send)

    # 如果是生命周期类型的请求,则由该程序段处理
    elif request_type == 'lifespan':
        await lifespan_handler.handle_lifespan(scope, receive, send)
    else:
        Cms4pyLog.get_instance().warning("Unsupported ASGI type")
```

对应的 error_pages.py 文件源码如下：

```
"""
第 8 章/cms4py_first_generation/cms4py/handlers/error_pages.py
"""


async def send_404_error(scope, receive, send):
    await send({
        'type': 'http.response.start',
        'status': 404,
        'headers': [
            [b'content-type', b'text/html'],
        ]
    })
    await send({
        'type': 'http.response.body',
        'body': b"Not found",
        'more_body': False
    })
```

对应的 lifespan_handler.py 文件源码如下：

```
"""
第 8 章/cms4py_first_generation/cms4py/handlers/lifespan_handler.py
"""
from cms4py.utils.log import Cms4pyLog


async def handle_lifespan(scope, receive, send):
    while True:
        # 不断读取数据
        message = await receive()
        # 如果读取消息类型为 lifespan.startup,则进行初始化操作
        if message['type'] == 'lifespan.startup':
            # 在初始化完成后,向 ASGI 环境发送启动完成消息
```

```
            await send({'type': 'lifespan.startup.complete'})
            Cms4pyLog.get_instance().info("Server started")
        # 如果读取消息类型为 lifespan.shutdown,则进行收尾工作
        elif message['type'] == 'lifespan.shutdown':
            # 在收尾工作结束后,向 ASGI 环境发送收尾完成消息
            await send({'type': 'lifespan.shutdown.complete'})
            Cms4pyLog.get_instance().info("Server stopped")
            break
```

在安装 uvicorn 之后可通过命令 uvicorn cms4py:application 启动服务器，如图 8-2 所示。

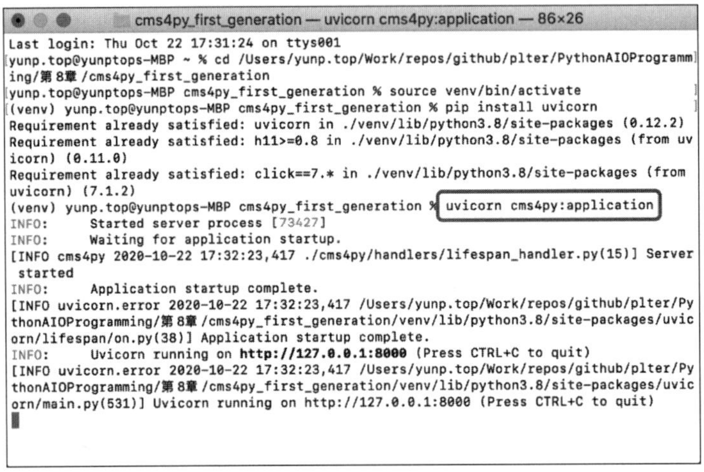

```
cms4py_first_generation — uvicorn cms4py:application — 86×26
Last login: Thu Oct 22 17:31:24 on ttys001
yunp.top@yunptops-MBP ~ % cd /Users/yunp.top/Work/repos/github/plter/PythonAIOProgramm
ing/第8章/cms4py_first_generation
yunp.top@yunptops-MBP cms4py_first_generation % source venv/bin/activate
(venv) yunp.top@yunptops-MBP cms4py_first_generation % pip install uvicorn
Requirement already satisfied: uvicorn in ./venv/lib/python3.8/site-packages (0.12.2)
Requirement already satisfied: h11>=0.8 in ./venv/lib/python3.8/site-packages (from uv
icorn) (0.11.0)
Requirement already satisfied: click==7.* in ./venv/lib/python3.8/site-packages (from
uvicorn) (7.1.2)
(venv) yunp.top@yunptops-MBP cms4py_first_generation % uvicorn cms4py:application
INFO:     Started server process [73427]
INFO:     Waiting for application startup.
[INFO cms4py 2020-10-22 17:32:23,417 ./cms4py/handlers/lifespan_handler.py(15)] Server
 started
INFO:     Application startup complete.
[INFO uvicorn.error 2020-10-22 17:32:23,417 /Users/yunp.top/Work/repos/github/plter/Py
thonAIOProgramming/第8章/cms4py_first_generation/venv/lib/python3.8/site-packages/uvic
orn/lifespan/on.py(38)] Application startup complete.
INFO:     Uvicorn running on http://127.0.0.1:8000 (Press CTRL+C to quit)
[INFO uvicorn.error 2020-10-22 17:32:23,417 /Users/yunp.top/Work/repos/github/plter/Py
thonAIOProgramming/第8章/cms4py_first_generation/venv/lib/python3.8/site-packages/uvic
orn/main.py(531)] Uvicorn running on http://127.0.0.1:8000 (Press CTRL+C to quit)
```

图 8-2 启动 cms4py 服务器

8.3 处理静态文件请求

处理静态文件请求的原理是根据所配置的静态文件根目录与浏览器请求的路径拼接出该文件在本机的真实路径，读取该文件的内容并将其返回浏览器端。

拼接文件真实路径的代码如下：

```
file_path = f"{config.STATIC_FILES_ROOT}{scope['path']}"
```

浏览器端需要通过 mime_type 来确定文件的类型，以选择合适的打开方式，所以服务器端应当提供文件后缀与 mime_type 的对应表，后缀与 mime_type 所有对应数据可参考 Mozilla 的 HTTP 协议文档，一个简单的示例代码如下：

```
mime_type_map = {
    ".html": b"text/html",
    ".htm": b"text/html",
    ".js": b"text/JavaScript",
    ".css": b"text/css",
```

```
    ".jpg": b"image/jpeg",
    ".jpeg": b"image/jpeg",
    ".png": b"image/png",
    ".gif": b"image/gif",
}
```

服务器通过文件的后缀获取对应的 mime_type，并以其配置 HTTP 协议头的 content-type 字段，完整的示例代码如下：

```
"""
第 8 章/cms4py_first_generation/cms4py/handlers/static_file_handler.py
"""

import config
from cms4py.utils import aiofile

#mime_type 表，这里列出最常用的文件类型，后期可继续完善
mime_type_map = {
    ".html": b"text/html",
    ".htm": b"text/html",
    ".js": b"text/JavaScript",
    ".css": b"text/css",
    ".jpg": b"image/jpeg",
    ".jpeg": b"image/jpeg",
    ".png": b"image/png",
    ".gif": b"image/gif",
}


def get_mime_type(file_path: str) -> Bytes:
    """
根据文件路径获取对应的 mime_type
    :param file_path: 文件路径
    :return: 对应的 mime_type
    """

    last_dot_index = file_path.rfind(".")
    #设定 mime_type 的默认值为 text/plain
    mime_type = b"text/plain"
    if last_dot_index > -1:
        #获取文件后缀名
        file_type = file_path[last_dot_index:]
        if file_type in mime_type_map:
            #根据后缀名获取 mime_type
            mime_type = mime_type_map[file_type]
    return mime_type


async def handle_static_file_request(scope, send) -> bool:
```

```
    """
处理静态文件请求
    :param scope:
    :param send:
    :return: 如果文件存在并且已经发送了数据,则返回 True,否则返回 False
    """

    # 因为对于静态文件的请求均为 GET 方式,所以其他方式的
    # 静态文件请求可视为非法请求,直接忽略即可
    if scope['method'] != 'GET':
        return False

    data_sent = False

    # 根据静态文件根目录拼接出该静态文件的绝对路径
    file_path = f"{config.STATIC_FILES_ROOT}{scope['path']}"
    file_content = None
    # 如果该路径存在并且存在文件,则读取该文件的内容并返回浏览器端
    if await aiofile.exists(file_path) and await aiofile.isfile(file_path):
        # 读取文件的二进制数据
        file_content = await aiofile.read_file(file_path)
    if file_content:
        # 获取文件对应的 mime type
        mime_type = get_mime_type(file_path)
        await send({
            'type': 'http.response.start',
            'status': 200,
            'headers': [
                [b'content-type', mime_type],
            ]
        })
        await send({
            'type': 'http.response.body',
            'body': file_content,
            'more_body': False
        })
        data_sent = True
    return data_sent
```

对应的 aiofile.py 文件代码如下：

```
"""
第 8 章/cms4py_first_generation/cms4py/utils/aiofile.py
"""

import asyncio, os
from typing import Any
```

```
class AsyncFunWrapper:

    def __init__(self, blocked_fun) -> None:
        super().__init__()

        # 记录阻塞型 IO 函数,便于后续调用
        self._blocked_fun = blocked_fun

    def __call__(self, *args):
        """
重载函数调用运算符,将阻塞型 IO 的调用过程异步化,并返回一个可
等待对象(Awaitable)。通过重载运算符实现包装逻辑的好处是不
用一个一个去实现阻塞型 IO 的所有成员函数,从而大大节省了代码
        """
        return asyncio.get_running_loop().run_in_executor(
            None,
            self._blocked_fun,
            *args
        )


class AIOWrapper:
    def __init__(self, blocked_file_io) -> None:
        super().__init__()
        # 在包装器对象中记录阻塞型 IO 对象,外界通过包装器调用其成员
        # 函数时,事实上是分为两步进行
        # 第一步,获取指定的成员,而该成员是一个可被调用的
        # 对象(Callable)
        # 第二步,对该成员进行调用
        self._blocked_file_io = blocked_file_io

    # 重载访问成员的运算符
    def __getattribute__(self, name: str) -> Any:
        """
在外界通过包装器(AIOWrapper)访问成员操作时,创建一个异步
函数包装器(AsyncFunWrapper),其目的是将函数调用过程异步化
        """
        return AsyncFunWrapper(
            super().__getattribute__(
                "_blocked_file_io"
            ).__getattribute__(name)
        )


async def open_async(*args) -> AIOWrapper:
    """
当外界调用该函数时,将返回一个包装器(AIOWrapper)对象,该包装器
包装了一个阻塞型 IO 对象
    """
    return AIOWrapper(
```

```
        #通过 run_in_executor 函数执行阻塞型 IO 的 open 函数,并转发外
        #界传入的参数
        await asyncio.get_running_loop().run_in_executor(
            None, open, *args
        )
    )


async def read_file(file_path) -> Bytes:
    """
以二进制模式读取文件内容
    :param file_path:
    :return:
    """
    f = await open_async(file_path, "rb")
    content = await f.read()
    await f.close()
    return content


async def exists(file_path) -> bool:
    """
判断指定的文件是否存在
    :param file_path:
    :return:
    """
    return await asyncio.get_running_loop().run_in_executor(
        None, os.path.exists, file_path
    )


async def isfile(file_path) -> bool:
    """
判断指定的路径是否存在文件
    :param file_path:
    :return:
    """
    return await asyncio.get_running_loop().run_in_executor(
        None, os.path.isfile, file_path
    )


async def getmtime(file_path):
    """
获取指定路径文件的修改时间
    :param file_path:
    :return:
    """
    return await asyncio.get_running_loop().run_in_executor(
        None, os.path.getmtime, file_path
    )
```

为实现静态文件请求处理功能，接下来将 __init__. py 文件的源码进行修改，修改后的代码如下：

```
"""
第 8 章/cms4py_first_generation/cms4py/__init__.py
"""

from cms4py.handlers import lifespan_handler
from cms4py.handlers import error_pages
from cms4py.utils.log import Cms4pyLog
from cms4py.handlers import static_file_handler


async def application(scope, receive, send):
    # 获取请求类型
    request_type = scope['type']

    # 如果是 http 类型的请求，则由该程序段处理
    if request_type == 'http':
        data_sent = await static_file_handler.handle_static_file_request(
            scope, send
        )

        # 如果静态文件处理程序未发送数据，则意味着找不到文件，此时应该
        # 向浏览器发送 404 页面
        if not data_sent:
            # 对于未被处理的请求，均向浏览器发回 404 错误
            await error_pages.send_404_error(scope, receive, send)

    # 如果是生命周期类型的请求，则由该程序段处理
    elif request_type == 'lifespan':
        await lifespan_handler.handle_lifespan(scope, receive, send)
    else:
        Cms4pyLog.get_instance().warning("Unsupported ASGI type")
```

由于所配置的静态文件目录为 app/static，所以需要在 app/static 目录创建一个 index.html 文件以便测试该功能是否能够正常运行。此时项目文件结构如图 8-3 所示。

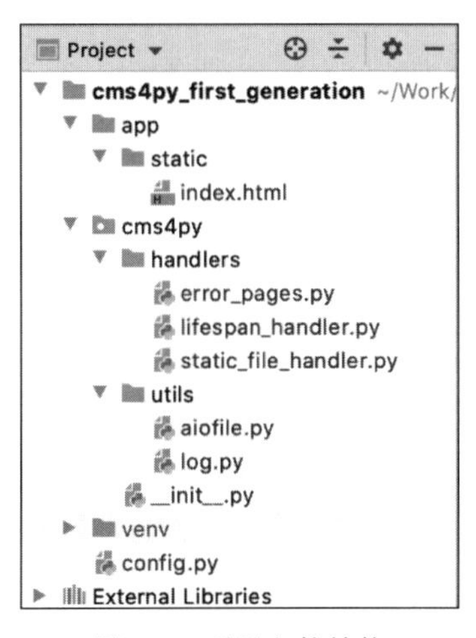

图 8-3　项目文件结构

其中 index.html 文件的内容如下：

```
<!-- 第 8 章/cms4py_first_generation/app/static/index.html -->
<!DOCTYPE html>
<html lang="en">
<head>
<meta charset="UTF-8">
<title>Title</title>
</head>
<body>
Hello cms4py
</body>
</html>
```

启动服务器后，在浏览器中访问 http://127.0.0.1:8000/index.html，效果如图 8-4 所示。

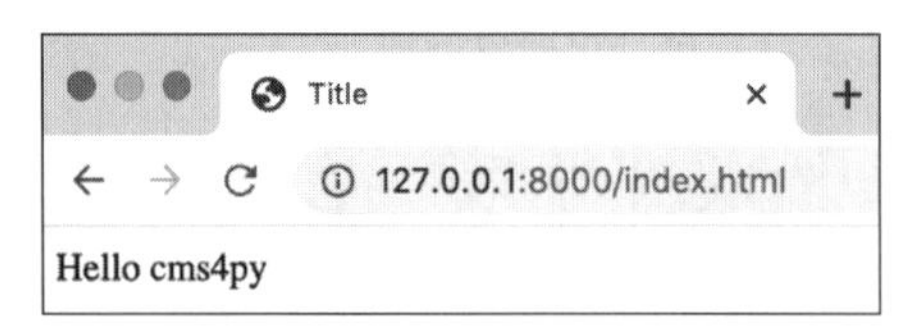

图 8-4　静态文件访问效果

8.4　静态文件缓存

笔者实现这个静态文件处理功能并非只是把它当作玩具练手用，更是希望能将它应用在生产环境中。在中小型网站的部署中，可以直接把 cms4py 当成与前端对接的入口，这样部署简单，不再需要依赖任何其他技术（如：Apache 或 Nginx）而可以独立运行。

在上一节完全通过异步 IO 实现文件的处理，就是为了给支持高并发进行铺垫，但这还不够。在 HTTP 中定义了缓存机制，可减少不必要的文件请求，以降低服务器端的压力，详细的缓存机制定义可参考 https://developer.mozilla.org/zh-CN/docs/Web/HTTP/Caching_FAQ。

千里之行始于足下，接下来从最核心的原理开始实现。原理就是服务器在发送文件数据的同时在响应头中指定该文件的最后修改时间，当浏览器请求同一文件时，会将该时间再次传给服务器，服务器可通过时间来对比该文件是否有更新，如果有更新，则重新读取文件数据并返回，否则向浏览器发送 304 状态码，浏览器在收到 304 状态码后直接从本地读取该文件数据。

修改后的 static_file_handler.py 文件代码如下：

```
"""
第 8 章/cms4py_first_generation/cms4py/handlers/static_file_handler.py
"""

import config, datetime
```

```
from cms4py.utils import aiofile
from cms4py.utils import http_helper

#mime_type 表,这里列出最常用的文件类型,后期可继续完善
mime_type_map = {
    ".html": b"text/html",
    ".htm": b"text/html",
    ".js": b"text/JavaScript",
    ".css": b"text/css",
    ".jpg": b"image/jpeg",
    ".jpeg": b"image/jpeg",
    ".png": b"image/png",
    ".gif": b"image/gif",
}


def get_mime_type(file_path: str) -> Bytes:
    """
根据文件路径获取对应的 mime_type
    :param file_path: 文件路径
    :return: 对应的 mime_type
    """

    last_dot_index = file_path.rfind(".")
    #设定 mime_type 的默认值为 text/plain
    mime_type = b"text/plain"
    if last_dot_index > -1:
        #获取文件后缀名
        file_type = file_path[last_dot_index:]
        if file_type in mime_type_map:
            #根据后缀名获取 mime_type
            mime_type = mime_type_map[file_type]
    return mime_type


async def handle_static_file_request(scope, send) -> bool:
    """
处理静态文件请求
    :param scope:
    :param send:
    :return: 如果文件存在并且已经发送了数据,则返回 True,否则返回 False
    """

    #因为对于静态文件的请求均为 GET 方式,所以其他方式的
    #静态文件请求可视为非法请求,直接忽略即可
    if scope['method'] != 'GET':
        return False

    data_sent = False
```

```
    #根据请求路径和静态文件根路径组装成文件真实路径
    file_path = f"{config.STATIC_FILES_ROOT}{scope['path']}"

    #如果指定路径存在并且存在文件,则读取文件的内容并向浏览器返回数据
    if await aiofile.exists(file_path) and await aiofile.isfile(file_path):
        #获取文件的 mime_type
        mime_type = get_mime_type(file_path)

        #获取文件的修改时间
        file_timestamp = datetime.datetime.utcfromtimestamp(
            await aiofile.getmtime(file_path)
        )
        #将文件的修改时间转换成 HTTP 协议标准的时间字符串
        file_timestamp_http_time_str = http_helper.datetime_to_http_time(
            file_timestamp
        )
        file_timestamp_http_time_Bytes: Bytes = file_timestamp_http_time_str.encode(
            config.GLOBAL_CHARSET
        )

        #读取浏览器端发来的请求头信息中的 if-modified-since 字段
        headers = scope['headers']
        if_modified_since_value_Bytes = b''
        if headers and len(headers):
            for h in headers:
                if len(h) >= 2:
                    hk = h[0]
                    hv = h[1]
                    if hk == b'if-modified-since':
                        if_modified_since_value_Bytes = hv
                        break

        if if_modified_since_value_Bytes:
            #如果时间相同,意味着文件未被改变,则向浏览器发送 304 状态码
            if if_modified_since_value_Bytes == file_timestamp_http_time_Bytes:
                #发送 304 状态码
                await send({
                    'type': 'http.response.start',
                    'status': 304,
                    'headers': [
                        [b'content-type', mime_type],
                        [b'last-modified', file_timestamp_http_time_Bytes]
                    ]
                })
                await send({
                    'type': 'http.response.body',
                    'body': b"",
                    'more_body': False
                })
```

```
            data_sent = True
        # 如果在前面的逻辑中没有发送数据,则意味着需要读取文件数据并发送给浏览器
        if not data_sent:
            await send({
                'type': 'http.response.start',
                'status': 200,
                'headers': [
                    [b'content-type', mime_type],
                    [b'last-modified', file_timestamp_http_time_Bytes]
                ]
            })
            file_content = await aiofile.read_file(file_path)
            await send({
                'type': 'http.response.body',
                'body': file_content,
                'more_body': False
            })
            data_sent = True
    return data_sent
```

对应的 http_helper.py 文件代码如下：

```
"""
第 8 章/cms4py_first_generation/cms4py/utils/http_helper.py
"""

import re, datetime

months = [
    "Jan", "Feb", "Mar", "Apr", "May", "Jun",
    "Jul", "Aug", "Sep", "Oct", "Nov", "Dec"
]
month_str_to_num_map = {
    b"Jan": 1, b"Feb": 2, b"Mar": 3, b"Apr": 4, b"May": 5, b"Jun": 6,
    b"Jul": 7, b"Aug": 8, b"Sep": 9, b"Oct": 10, b"Nov": 11, b"Dec": 12
}
week_days = ["Mon", "Tue", "Wed", "Thu", "Fri", "Sat", "Sun"]


def format_time(tn) -> str:
    """
在个位数前面补 0
    :param tn:
    :return:
    """
    return f'{tn}' if tn >= 10 else f"0{tn}"


def datetime_to_http_time(dt) -> str:
    """
将 Python 的 datetime 转换为 HTTP 协议标准的时间字符串
    :param dt:
```

```
    :return:
    """
    return f"{week_days[dt.weekday()]}, " \
           f"{format_time(dt.day)} {months[dt.month - 1]} {dt.year} " \
           f"{format_time(dt.hour)}:{format_time(dt.minute)}:{format_time(dt.second)} " \
           f"GMT"


def http_time_to_datetime(http_time: Bytes):
    """
将 HTTP 协议标准的时间字符串转成 Python 的 datetime
注意:时间字符串为 Bytes 类型
    :param http_time:
    :return:
    """

    #使用正则表达式匹配时间字符串,并截取其中的年、月、日信息
    result = re.match(
        b'\\w{3}, (\\d{2}) (\\w{3}) (\\d{4}) (\\d{2}):(\\d{2}):(\\d{2}) GMT',
        http_time
    )
    if result:
        #根据年、月、日信息组装成一个 datetime
        day = int(result.group(1))
        month = month_str_to_num_map[result.group(2)]
        year = int(result.group(3))
        hour = int(result.group(4))
        minute = int(result.group(5))
        second = int(result.group(6))
        return datetime.datetime(
            year = year, month = month, day = day,
            hour = hour, minute = minute, second = second
        )
    return None
```

图 8-5 为初次访问与再次访问的效果对比。

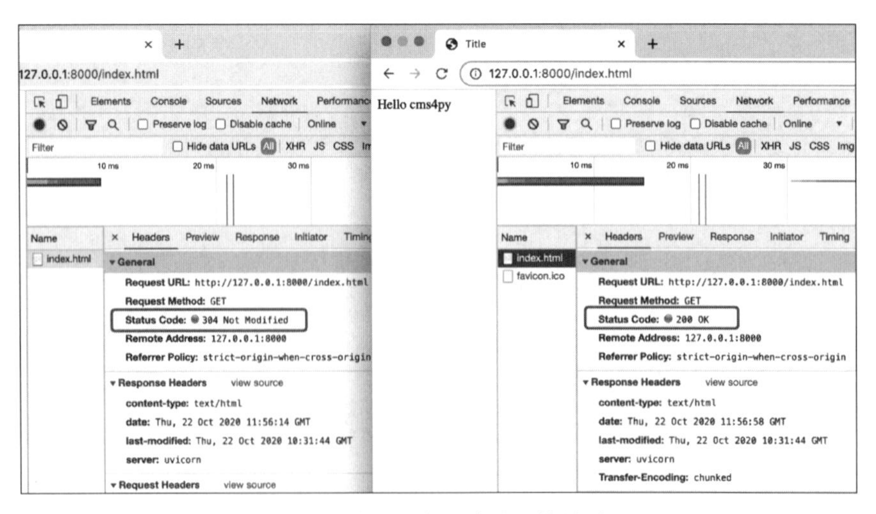

图 8-5 初次访问与再次访问效果对比

8.5 处理动态请求

在 cms4py 中，请求路径被设定为与控制器对应，cms4py 将请求路径分拆为控制器(Controller)和动作(Action)进行处理。例如：浏览器对路径/user/list_all 请求对应的 Controller 是 user，而 Action 是 list_all。这样做可直接使请求与处理文件对应，其好处是免去了配置路由的麻烦，从而大大提高开发效率。

动态页面的请求处理器源码如下：

```
"""
第 8 章/cms4py_first_generation/cms4py/handlers/dynamic_handler.py
"""

import config
import os, inspect
import importlib.util
from cms4py.utils import aiofile
from cms4py import http


async def handle_dynamic_request(scope, receive, send) -> bool:
    data_sent = False
    request_path: str = scope['path']

    # 将请求路径分拆为 Controller 和 Action，例如:/user/list_all 请求
    # 对应的 Controller 是 user，而 Action 是 list_all
    tokens = request_path.split("/")
    tokens_len = len(tokens)
    # 指定默认的控制器名
    controller_name = config.DEFAULT_CONTROLLER
    if tokens_len >= 2:
        controller_name = tokens[1] or config.DEFAULT_CONTROLLER
    # 指定默认的函数名
    action_name = config.DEFAULT_ACTION
    if tokens_len >= 3:
        action_name = tokens[2] or config.DEFAULT_ACTION
    controller_file = os.path.join(
        config.CONTROLLERS_ROOT, f"{controller_name}.py"
    )

    controller_object = None
    # 如果文件存在，则尝试将该文件导入为模块
    if await aiofile.exists(controller_file):
        # 根据浏览器请求路径导入指定的模块
        controller_object = importlib.import_module(
            f"{config.APP_DIR_NAME}.{config.CONTROLLERS_DIR_NAME}.{controller_name}"
        )
```

```
            pass

    if controller_object:
        #根据 action_name 获取指定的成员
        action = getattr(controller_object, action_name, None)
        if action:
            #构造 HTTP 请求对象,便于后续操作
            req = http.Request(scope, receive)
            #构造 HTTP 响应对象,便于后续操作
            res = http.Response(req, send)
            req._controller = controller_name
            req._action = action_name
            #如果 action 是类定义,则先将类实例化再执行
            if inspect.isclass(action):
                await action()(req, res)
            #否则把 action 当作函数对待并直接执行
            else:
                await action(req, res)
            data_sent = res.body_sent
    return data_sent
```

对应的 http 模块初始化文件代码如下：

```
"""
第 8 章/cms4py_first_generation/cms4py/http/__init__.py
"""
from cms4py.http.response import Response
from cms4py.http.request import Request
```

对应的 request.py 文件代码如下：

```
"""
第 8 章/cms4py_first_generation/cms4py/http/request.py
"""

import config
import re


class Request:
    """
将浏览器的请求封装为一个 Request 对象,便于操作
    """

    def __init__(self, scope, receive):
        self._scope = scope
        self._receive = receive
        #记录协议类型, http 或者 https
        self._protocol = scope['scheme']
```

```
        # 记录请求方法
        self._method = self._scope['method']
        # 记录请求路径
        self._path = self._scope['path']
        # 记录请求路径中 ? 后面的字符串
        self._query_string = self._scope['query_string']
        # 记录 ASGI 发来的原始请求头数据
        self._raw_headers = self._scope['headers'] \
            if 'headers' in self._scope else []
        # 声明一个字典,用于存放头数据
        self._headers = {}
        # 将原始请求头数据封装进 self._headers,便于后续使用
        self._copy_headers()
        # 记录原始请求头中的可接收的语言列表数据
        self._raw_accept_languages = self.get_header(b'accept-language')
        # 用正则表达将可接收的语言截取出来并转换成数组
        self._accept_languages = re.compile(b"[a-z]{2}-[A-Z]{2}").findall(
            self._raw_accept_languages
        ) if self._raw_accept_languages else []

        lang: Bytes = config.LANGUAGE or (
            self._accept_languages[0] if len(self._accept_languages) > 0 else b'en-US'
        )
        # 记录将要使用的语言种类
        self._language = lang.decode("utf-8")

        # 该变量用于记录 content_type
        self._content_type = None
        # 该变量用于记录请求的 uri
        self._uri = None
        # 记录请求的主机
        self._host = self.get_header(b"host")
        # 记录信息
        self._client = self._scope['client']
        # 记录客户端 ip 地址
        self._client_ip = self._client[0]
        # 记录客户端端口
        self._client_port = self._client[1]

        # 该变量用于记录请求的控制器名
        self._controller = None
        # 该变量用于记录请求的函数名
        self._action = None
        pass

    @property
    def controller(self):
        return self._controller

    @property
```

```
    def action(self):
        return self._action

    @property
    def host(self) -> Bytes:
        return self._host

    def host_as_str(self, charset = config.GLOBAL_CHARSET) -> str:
        return self.host.decode(charset) if self.host else ''

    @property
    def client_ip(self):
        return self._client_ip

    @property
    def protocol(self) -> str:
        return self._protocol

    @property
    def uri(self) -> str:
        if not self._uri:
            self._uri = self.path
            if self.query_string:
                self._uri += "?"
                self._uri += self.query_string.decode(
                    config.GLOBAL_CHARSET
                )
        return self._uri

    def _copy_headers(self):
        #将原始请求头数据封装进 self._headers
        #在 HTTP 协议中,消息头支持多条同名的字段,所以字段名
        #对应的是列表对象
        for pair in self._raw_headers:
            if len(pair) == 2:
                key = pair[0]
                if key not in self._headers:
                    self._headers[key] = []
                self._headers[key].append(pair[1])
        pass

    @property
    def language(self) -> str:
        return self._language

    @property
    def accept_languages(self):
        return self._accept_languages

    @property
```

```
    def headers(self):
        return self._headers

    def is_mobile(self):
        user_agent = self.get_header(b"user-agent")
        if user_agent:
            return user_agent.find(b"iPhone") != -1 or \
                   user_agent.find(b"iPad") != -1 or \
                   user_agent.find(b"Android") != -1
        return False

    @property
    def query_string(self) -> Bytes:
        return self._query_string

    def _get_first_value_of_array_map(self, data, key):
        values = data[key] if (data and key in data) else None
        value = None
        if values and len(values) > 0:
            value = values[0]
        return value

    def get_headers(self, key: Bytes):
        """
        Get all values by key
        :param key:
        :return:
        """
        return self.headers[key] if key in self.headers else None

    def get_header(self, key: Bytes, default_value=None) -> Bytes:
        """
        Get first value by key
        :param key:
        :param default_value:
        :return:
        """
        return self._get_first_value_of_array_map(
            self.headers, key
        ) or default_value

    @property
    def content_type(self) -> Bytes:
        if not self._content_type:
            self._content_type = self.get_header(b"content-type")
        return self._content_type

    @property
    def method(self) -> str:
        return self._method
```

```
    @property
    def path(self) -> str:
        return self._path

    pass
```

对应的 response.py 文件代码如下：

```
"""
第 8 章/cms4py_first_generation/cms4py/http/response.py
"""

import config
from cms4py.http.request import Request


class Response:
    def __init__(self, request: Request, send):
        self._send = send
        #该变量用于记录返回的 content_type 类型
        self._content_type = None
        #该变量用于指示头部是否已发送
        self._header_sent = False
        #该变量用于指示内容是否已发送
        self._body_sent = False
        self._body = b''
        #记录 Request 对象
        self._request: Request = request

        #返回的头部信息
        self._headers_map = {}

        #指定默认的 content_type 是 text/html
        self.content_type = b'text/html'
        #添加自定义的服务器名称信息
        self.add_header(b'server', config.SERVER_NAME)
        pass

    @property
    def header_sent(self):
        return self._header_sent

    def _get_headers(self):
        result = []
        for key in self._headers_map:
            for v in self._headers_map[key]:
                result.append([key, v])
        return result
```

```
    def add_header(self, key: Bytes, value: Bytes):
        if key not in self._headers_map:
            self._headers_map[key] = []
        self._headers_map[key].append(value)

    @property
    def body_sent(self):
        return self._body_sent

    @property
    def content_type(self) -> Bytes:
        return self._content_type

    @content_type.setter
    def content_type(self, value: Bytes):
        self._content_type = value
        self._headers_map[b"content-type"] = [value]

    @property
    def body(self):
        return self._body

    async def send_header(self, status=200):
        """
发送协议头
        :param status:
        :return:
        """
        await self._send({
            'type': 'http.response.start',
            'status': status,
            'headers': self._get_headers()
        })
        self._header_sent = True

    async def write(self, data: Bytes):
        """
向浏览器端写数据,但不关闭连接
        :param data:
        :return:
        """
        if not self._header_sent:
            await self.send_header()
        await self._send({
            "type": "http.response.body",
            "body": data,
            # 如果指定 more_body 为 True,则意味着还有待发送的
            # 数据,连接需要被继续保持
            'more_body': True
        })

    async def end(self, data: Bytes):
        """
```

```
该函数用于发送数据之后关闭连接
        :param data:
        :return:
        """
        if self._body_sent:
            return
        if not self._header_sent:
            await self.send_header()
        await self._send({
            "type": "http.response.body",
            "body": data,
            'more_body': False
        })
        self._body = data
        self._body_sent = True
```

接下来需要更新 cms4py 的初始文件__init__.py 以连接动态页面处理功能，对其源码进行修改，修改后的代码如下：

```
"""
第 8 章/cms4py_first_generation/cms4py/__init__.py
"""

from cms4py.handlers import lifespan_handler
from cms4py.handlers import error_pages
from cms4py.utils.log import Cms4pyLog
from cms4py.handlers import static_file_handler
from cms4py.handlers import dynamic_handler


async def application(scope, receive, send):
    # 获取请求类型
    request_type = scope['type']

    # 如果是 http 类型的请求,则由该程序段处理
    if request_type == 'http':
        data_sent = await dynamic_handler.handle_dynamic_request(
            scope, receive, send
        )

        # 如果经过动态请求处理程序后未发送数据,则尝试使用静态文件请求
        # 处理程序处理
        if not data_sent:
            data_sent = await static_file_handler.handle_static_file_request(
                scope, send
            )

        # 如果静态文件处理程序未发送数据,则意味着文件找不到,此时应该
        # 向浏览器发送 404 页面
```

```
        if not data_sent:
            # 对于未被处理的请求,均向浏览器发送 404 错误
            await error_pages.send_404_error(scope, receive, send)

    # 如果是生命周期类型的请求,则由该程序段处理
    elif request_type == 'lifespan':
        await lifespan_handler.handle_lifespan(scope, receive, send)
    else:
        Cms4pyLog.get_instance().warning("Unsupported ASGI type")
```

根据配置需要在 app 目录下创建 controllers 目录,并在 controllers 目录下创建一个 default.py 文件,此时项目文件结构如图 8-6 所示。

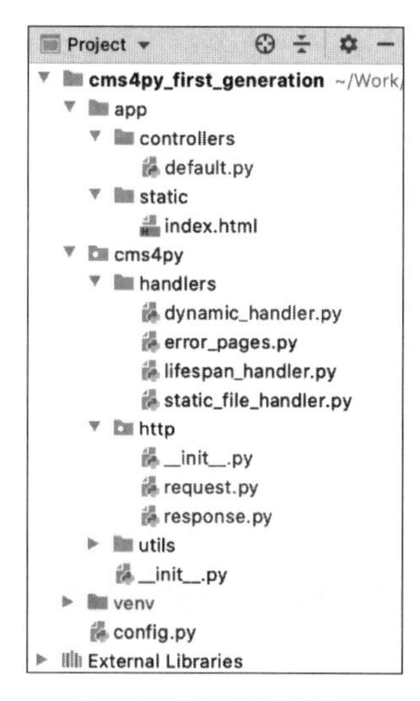

图 8-6　项目文件结构

文件 default.py 的源码如下:

```
"""
第 8 章/cms4py_first_generation/app/controllers/default.py
"""


async def index(req, res):
    await res.end(b"Home page")
    pass


class hello:
    """
如果 action 是类,则需要使用魔术函数 __call__ 重载函数调用
运算符以接受调用操作
    """
    async def __call__(self, req, res):
        await res.end(b"Hello cms4py")
```

在启动服务器后，访问这两个 action 的效果如图 8-7 所示。

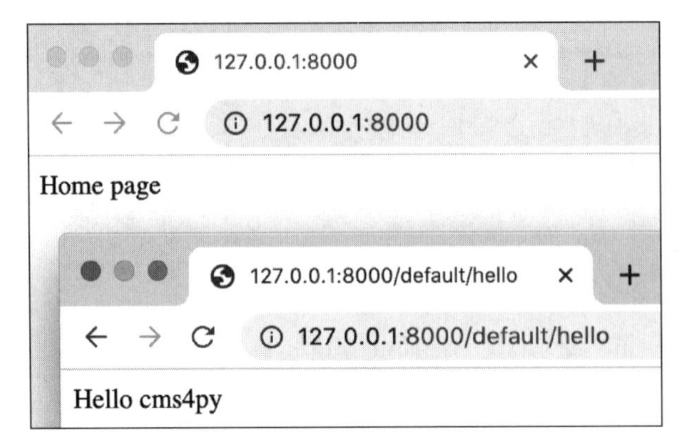

图 8-7　访问两个 action 的效果

8.6　实现控制器热更新

在 8.5 节中，使用 importlib.import_module 函数来动态加载模块，但默认情况下，一个模块被加载后，它将被缓存起来，就算该模块的源码被改变了，重新执行该模块仍然不能导入最新的代码，所以当一个控制器被导入后，再修改控制器中的代码，刷新页面还是只能看到修改前的效果，若要使修改生效需要重启服务器。

在实际开发工作中，如果因为小改动而去重启服务器，显得不值。正常运行中的服务器，如果有在线交易服务，重启则意味着随时都会有损失。如果 cms4py 框架能够提供一个高效的热更新机制，那将是非常完美的事。

这就需要使用 importlib.reload 函数重新加载该模块，在应用最新代码的同时无须重启服务器。但如果每一次请求都重新加载模块，又是一种巨大的资源浪费，因为重新加载意味着要重新读取文件内容并进行编译，这是个很耗时的过程，这就要求我们设计一个完美的缓存管理机制，从而实现在文件内容没有修改的情况下不重新加载模块。

因为在服务器开发工作中缓存功能是很常用的，例如：页面缓存、模板缓存、语言缓存等，所以笔者倾向于设计一个统一的缓存机制，从而最大限度地复用代码。为此笔者实现一个统一的抽象父类，源码如下：

```
"""
第 8 章/cms4py_first_generation/cms4py/cache/base_cache_manager.py
"""


class CachedDataWrapper:
    """
该类用于包装缓存的数据
    """

    def __init__(self, data, timestamp) -> None:
```

```
        super().__init__()
        # 记录数据缓存的时间戳
        self.timestamp = timestamp
        # 记录缓存的数据
        self.data = data


class BaseCacheManager:
    """
缓存管理器抽象基类,该类仅实现缓存管理功能,至于何时
更新缓存、缓存什么数据,交由子类实现
    """

    def __init__(self) -> None:
        super().__init__()
        # 建立一个字典,用于缓存数据
        self._cache_map = {}
        # 实现包装数据的回调函数
        self._wrap_data_callback = None
        # 实现确定何时重新缓存数据的回调函数
        self._will_reload_callback = None

    async def wrap_data(self, key) -> CachedDataWrapper:
        """
创建包装的数据,返回值将由缓存机制进行缓存,该函数应由子
类实现
        :param key:
        :return:
        """
        raise NotImplementedError()

    async def retrieve_cache_data(self, key):
        """
根据键名直接从缓存中取数据
        :param key:
        :return:
        """
        if not self._wrap_data_callback:
            self._wrap_data_callback = self.wrap_data

        wrapper = await self._wrap_data_callback(key)
        if wrapper:
            self._cache_map[key] = wrapper
        return wrapper.data if wrapper else None

    async def get_data(
            self, key,
            wrap_data_callback = None, will_reload_callback = None
    ):
        """
```

```
获取数据,该函数将自动根据需要创建新数据或返回缓存的数据
        :param key: 缓存键
        :param wrap_data_callback: 包装数据的回调函数,置空将调用
                                   self.wrap_data 函数以创建数据
        :param will_reload_callback: 提示是否创建新数据的回调函数,置空
                                     将使用 self.will_reload 进行回调
        :return:
        """

        self._wrap_data_callback = wrap_data_callback or self.wrap_data
        self._will_reload_callback = will_reload_callback or self.will_reload

        if key in self._cache_map:
            wrapper: CachedDataWrapper = self._cache_map[key]
            result = wrapper.data
            if await self._will_reload_callback(wrapper, key):
                result = await self.retrieve_cache_data(key)
        else:
            result = await self.retrieve_cache_data(key)
        return result

    async def will_reload(self, wrapper: CachedDataWrapper, key: str) -> bool:
        """
子类重写该函数以指示是否需要创建新数据
        :param wrapper:
        :param key:
        :return:
        """
        raise NotImplementedError()

    def clear(self):
        """
清除所有缓存
        :return:
        """
        self._cache_map = {}

    @property
    def cache_map(self):
        return self._cache_map
```

模块缓存管理类继承该抽象父类并实现导入和重新加载模块功能,代码如下:

```
"""
第 8 章/cms4py_first_generation/cms4py/cache/modules_cache_manager.py
"""

from cms4py.cache.base_cache_manager import BaseCacheManager
from cms4py.cache.base_cache_manager import CachedDataWrapper
```

```
from cms4py.utils import aiofile
import importlib, config, os
from cms4py.utils.log import Cms4pyLog


class ModulesCacheManager(BaseCacheManager):
    __instance = None

    @staticmethod
    def get_instance() -> "ModulesCacheManager":
        if not ModulesCacheManager.__instance:
            ModulesCacheManager.__instance = ModulesCacheManager()
        return ModulesCacheManager.__instance

    def file_name_from_module_name(self, module_name):
        """
将模块名转换文件路径
        :param module_name:
        :return:
        """
        file_path_tokens = module_name.split(".")
        return os.path.join(
            config.SERVER_ROOT,
            file_path_tokens[0],
            file_path_tokens[1],
            f"{file_path_tokens[2]}.py"
        )

    async def wrap_data(self, key) -> CachedDataWrapper:
        m = None
        t = 0
        try:
            # 导入模块
            m = importlib.import_module(key)
            m_file = self.file_name_from_module_name(key)
            # 获取文件时间戳
            t = await aiofile.getmtime(m_file)
        except ModuleNotFoundError:
            Cms4pyLog.get_instance().warning(f"Module {key} not found")
        return CachedDataWrapper(m, t) if m else None

    async def will_reload(self, wrapper: CachedDataWrapper, key: str) -> bool:
        return_value = False
        if key in self.cache_map:
            m_file = self.file_name_from_module_name(key)
            # 如果已缓存的文件时间戳和现有文件时间戳不一致,则重新加载模块
            if await aiofile.exists(m_file) and \
                    await aiofile.isfile(m_file) and \
                    (await aiofile.getmtime(m_file)) != wrapper.timestamp:
                importlib.reload(wrapper.data)
```

```
                return_value = True
        else:
            return_value = True
        return return_value
```

目录 cache 的初始文件__init__.py 内容如下：

```
"""
第 8 章/cms4py_first_generation/cms4py/cache/__init__.py
"""

from cms4py.cache.base_cache_manager import CachedDataWrapper
from cms4py.cache.base_cache_manager import BaseCacheManager
from cms4py.cache.modules_cache_manager import ModulesCacheManager
```

若要对接该缓存管理功能，须对 dynamic_handler.py 文件内容进行修改，修改后的代码如下：

```
"""
第 8 章/cms4py_first_generation/cms4py/handlers/dynamic_handler.py
"""

import inspect

import config
from cms4py import http
from cms4py.cache import ModulesCacheManager


async def handle_dynamic_request(scope, receive, send) -> bool:
    data_sent = False
    request_path: str = scope['path']

    # 将请求路径分拆为 Controller 和 Action，例如：/user/list_all 请求
    # 对应的 Controller 是 user，而 Action 是 list_all
    tokens = request_path.split("/")
    tokens_len = len(tokens)
    # 指定默认的控制器名
    controller_name = config.DEFAULT_CONTROLLER
    if tokens_len >= 2:
        controller_name = tokens[1] or config.DEFAULT_CONTROLLER
    # 指定默认的函数名
    action_name = config.DEFAULT_ACTION
    if tokens_len >= 3:
        action_name = tokens[2] or config.DEFAULT_ACTION

    controller_object = await ModulesCacheManager.get_instance().get_data(
        f"{config.APP_DIR_NAME}.{config.CONTROLLERS_DIR_NAME}.{controller_name}"
    )
```

```
    if controller_object:
        # 根据 action_name 获取指定的成员
        action = getattr(controller_object, action_name, None)
        if action:
            # 构造 HTTP 请求对象,便于后续操作
            req = http.Request(scope, receive)
            # 构造 HTTP 响应对象,便于后续操作
            res = http.Response(req, send)
            req._controller = controller_name
            req._action = action_name
            # 如果 action 是类定义,则先将类实例化再执行
            if inspect.isclass(action):
                await action()(req, res)
            # 否则把 action 当作函数对待并直接执行
            else:
                await action(req, res)
            data_sent = res.body_sent
    return data_sent
```

此时项目文件结构如图 8-8 所示。

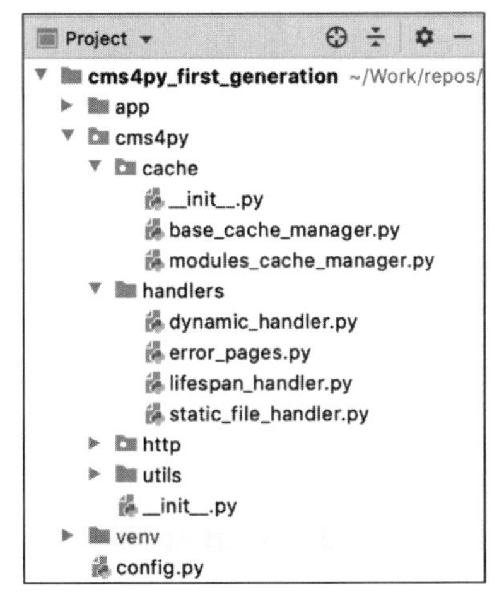

图 8-8　项目文件结构

8.7　实现动态页面缓存

一般来讲,动态页面需要与数据库进行交互,但有些页面从创建成功之后就再也不会改动。例如:博客页面、新闻稿页面、企业宣传页面等,如果每一次访问这些固定页面都查询数据库,将是巨大的资源浪费。

动态页面缓存技术的原理是将页面的返回数据记录在一个字典中,当再次访问同一页面时,直接从字典中取出数据返回给浏览器端,这样就避免了再次执行 action 而产生的数据库交互等一切不必要的资源浪费。在实际开发需求中,每个页面的更新频率可能都不同,所以在设计页面缓存机制时应当提供针对页面配置缓存过期时间的功能,同时 API 的设计必

须保证此用法的代码简单。

受 web2py 框架的启发，笔者计划将缓存机制的 API 用法设计为只需用一个装饰器，能够支持的 action 可以是函数和类，如果 action 是函数，则此用法的代码如下：

```
@cache(expire = 5)
async def index(req, res):
    await res.end(b"Home page")
```

如果 action 是类，则此用法的代码如下：

```
class hello:
    @cache(expire = 5)
    async def __call__(self, req, res):
        await res.end(b"Hello cms4py")
```

页面缓存的核心类源码如下：

```
"""
第 8 章/cms4py_first_generation/cms4py/cache/page_cache_manager.py
"""

from cms4py.cache.base_cache_manager import BaseCacheManager
from cms4py.cache.base_cache_manager import CachedDataWrapper
import datetime


class PageCacheManager(BaseCacheManager):
    __instance = None

    @staticmethod
    def get_instance():
        if not PageCacheManager.__instance:
            PageCacheManager.__instance = PageCacheManager()
        return PageCacheManager.__instance

    async def wrap_data(self, key) -> CachedDataWrapper:
        # 由于包装数据功能自定义程度较高，在该类内部难以获得所有必要数据
        # 所以该函数功能将由回调函数实现
        raise NotImplementedError()

    async def will_reload(self, wrapper: CachedDataWrapper, key: str) -> bool:
        # 当缓存的数据过期时重新加载数据
        return datetime.datetime.now().timestamp() > wrapper.timestamp
```

与之对应的装饰器源码如下：

```
"""
第 8 章/cms4py_first_generation/cms4py/cache/__init__.py
"""
```

```
from cms4py.cache.base_cache_manager import CachedDataWrapper
from cms4py.cache.base_cache_manager import BaseCacheManager
from cms4py.cache.modules_cache_manager import ModulesCacheManager
from cms4py.cache.page_cache_manager import PageCacheManager

from cms4py.utils.log import Cms4pyLog
import datetime


def cache(expire = 3600, key = None):
    """
为期望被缓存的页面添加该装饰器
    :param expire: 缓存过期时长,以秒为单位
    :param key: 缓存键
    :return:
    """

    def wrapper(target):

        async def inner( * args):
            argc = len(args)
            #如果参数的个数为 2,则目标 action 是函数
            if argc == 2:
                req = args[0]
                res = args[1]
            #如果参数的个数为 3,则目标 action 是类实例
            elif argc == 3:
                req = args[1]
                res = args[2]
            else:
                raise TypeError("Require 2 or 3 arguments")

            _key = key
            #如果没有指定键,则使用 uri 作为缓存键
            if not _key:
                _key = req.uri

                #移动端单独缓存
                if req.is_mobile():
                    _key += ",mobile"

            async def wrap_data_callback(cache_key) -> CachedDataWrapper:
                await target( * args)

                Cms4pyLog.get_instance().debug(f"Cache page {cache_key}")
                return CachedDataWrapper(
                    res.body,
                    #记录缓存过期时间,便于对比
                    datetime.datetime.now().timestamp() + expire
```

```
            )

            cached_data = await PageCacheManager.get_instance().get_data(
                _key, wrap_data_callback
            )
            if not res.body_sent:
                await res.end(cached_data)

        return inner

    return wrapper
```

在控制器中使用缓存功能的示例代码如下：

```
"""
第 8 章/cms4py_first_generation/app/controllers/default.py
"""
from cms4py.cache import cache


@cache(expire = 5)
async def index(req, res):
    await res.end(b"Home page")
    pass


class hello:
    """
如果 action 是类,则需要使用魔术函数 __call__ 重载函数调用
运算符以接受调用操作
    """

    @cache()
    async def __call__(self, req, res):
        await res.end(b"Hello cms4py")
```

8.8 实现路径参数解析功能

在 cms4py 框架中，将路径以/拆分，拆分出的列表可称为路径参数列表，其中索引 1 的值被当作控制器(Controller)，索引 2 的值被当作动作(Action)。在实际项目中，路径往往会很长，而更多字段可代表更多的含义，cms4py 应该提供获取更多路径参数的功能。

在 handle_dynamic_request 函数中添加如下代码以在 request 中记录路径参数：

```
req._args = tokens[3:] if len(tokens) > 3 else []
```

完成的 handle_dynamic_request 函数源码如下：

```
"""
第 8 章/cms4py_first_generation/cms4py/handlers/dynamic_handler.py
"""

import inspect

import config
from cms4py import http
from cms4py.cache import ModulesCacheManager


async def handle_dynamic_request(scope, receive, send) -> bool:
    data_sent = False
    request_path: str = scope['path']

    # 将请求路径分拆为 Controller 和 Action,例如:/user/list_all 请求
    # 对应的 Controller 是 user, 而 Action 是 list_all
    tokens = request_path.split("/")
    tokens_len = len(tokens)
    # 指定默认的控制器名
    controller_name = config.DEFAULT_CONTROLLER
    if tokens_len >= 2:
        controller_name = tokens[1] or config.DEFAULT_CONTROLLER
    # 指定默认的函数名
    action_name = config.DEFAULT_ACTION
    if tokens_len >= 3:
        action_name = tokens[2] or config.DEFAULT_ACTION

    controller_object = await ModulesCacheManager.get_instance().get_data(
        f"{config.APP_DIR_NAME}.{config.CONTROLLERS_DIR_NAME}.{controller_name}"
    )

    if controller_object:
        # 根据 action_name 获取指定的成员
        action = getattr(controller_object, action_name, None)
        if action:
            # 构造 HTTP 请求对象,便于后续操作
            req = http.Request(scope, receive)
            # 构造 HTTP 响应对象,便于后续操作
            res = http.Response(req, send)
            req._controller = controller_name
            req._action = action_name

            # 将 action 后的路径参数记录在 req.args 中
            req._args = tokens[3:] if len(tokens) > 3 else []
            # 如果 action 是类定义,则先将类实例化再执行
            if inspect.isclass(action):
                await action()(req, res)
```

```
            # 否则把 action 当作函数对待并直接执行
            else:
                await action(req, res)
            data_sent = res.body_sent
    return data_sent
```

在 Request 类中添加两个函数作为 API 提供给 Controller 使用,代码如下:

```
"""
第 8 章/cms4py_first_generation/cms4py/http/request.py
"""

import config
import re


class Request:
    """
将浏览器的请求封装为一个 Request 对象,便于操作
    """

    # >>>>>>>>>>>>
    # 提醒:此处省略部分重复的代码
    # <<<<<<<<<<<<

    @property
    def args(self):
        """
获取所有的路径参数
        :return:
        """
        return self._args

    def arg(self, index):
        """
返回指定位置的值或者 None
        :param index:
        :return:
        """
        return self.args[index] \
            if self.args and len(self.args) > index \
            else None
```

在 Controller 中使用该 API 的示例代码如下:

```
async def get_args(req, res):
    arg0 = req.arg(0)
    await res.write(b"Arg0 is ")
    await res.end(arg0.encode("utf-8") if arg0 else b"None")
```

在浏览器中运行的效果如图 8-9 所示。

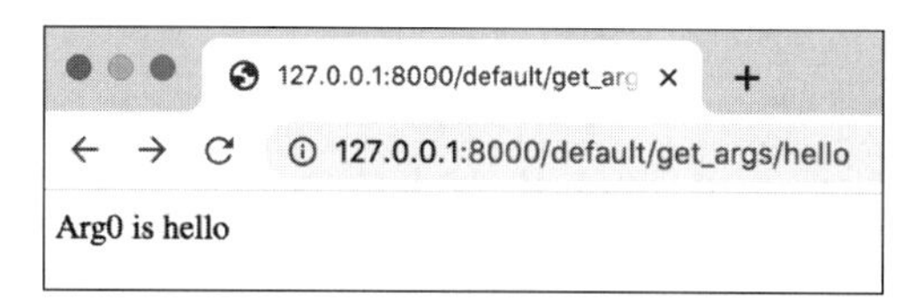

图 8-9　路径参数示例效果

8.9　实现表单解析功能

表单解析主要分为 GET 方式的表单解析与 POST 方式的表单解析，其中 POST 方式的表单解析还需要处理文件上传请求。本节难度较大，请先喝杯茶静静心。

从学习第 8 章以来，也许读者已经明显感觉到，这绝对不是在写用来玩的代码。每个人做任何事情的耐心都是有限的，不知道此时你是否还能坚持。但笔者要说，坚持住，你一定可以获得丰厚的回报。

浏览器上传的数据主要有 3 种编码方式，分别为 application/x-www-form-URLencoded、multipart/form-data 和 text/plain。

如果编码方式是 application/x-www-form-URLencoded，则数据的形式如 name=yunp&age=20，通常用于 GET 方式的表单提交，会出现在地址栏中的网页地址字符串中的问号后面，如 http://yunp.top?p=1&comment_index=2。在提交用户登录表单时，有时为了安全考虑，会以 POST 方式提交数据，此时数据出现在 HTTP 请求体中。解析这种编码方式的数据代码如下：

```
"""
第 8 章/cms4py_first_generation/cms4py/utils/URL_helper.py
"""


def parse_URL_pairs(query_string: Bytes):
    """
将 URL 字符串解码为字典,因为可能存在多条同键名的数据,
所以每个字段所对应的是一个列表
    :param query_string:
    :return:
    """

    params = {}

    #URL 字符串是以 & 符号连接的,如:name = yunp&age = 20
    #以逗号分割出的字符串数组如:[name = yunp,age = 20]
    #之后再用等号分别分割每个字符串,即可得到键值对
    tokens = query_string.split(b"&")
    for t in tokens:
```

```
            kv = t.split(b"=")
            if len(kv) == 2:
                k = kv[0]
                v = kv[1]
                if k not in params:
                    params[k] = []
                params[k].append(v)
    return params
```

如果编码方式是 multipart/form-data，则通常以 POST 方式向服务器发送数据，解析代码如下：

```
"""
第8章/cms4py_first_generation/cms4py/http/request.py
"""

#说明:此处省略部分代码

async def _parse_form(self):
    """
解析表单,该函数由 cms4py 框架内部调用,应用层不应该调用此函数
    :return:
"""

#路径中的参数
self._query_vars = {}
#协议内容中的参数
self._body_vars = {}

    #TODO 目前仅实现支持 GET 方式与 POST 方式,需要逐步完善
    #以支持所有的 HTTP 方式
    if self.query_string:
        #尝试从 URL 中解析参数
        self._query_vars = URL_helper.parse_URL_pairs(
            self.query_string
        )
    if self.method == "POST":
        #如果是 POST 方式,则需要尝试读取消息体
        while True:
            message = await self._receive()
            #TODO 需要实现数据限制机制以防攻击
            self._body += message["body"] if 'body' in message else b''
            if "more_body" not in message or not message["more_body"]:
                break
        if self.content_type:
            #如果是 application/x-www-form-URLencoded 编码方式
            #则尝试以 URL 参数对的方式解析
            if self.content_type.startswith(
                    b'application/x-www-form-URLencoded'
```

```
):
    self._body_vars = URL_helper.parse_URL_pairs(
        self._body
    )
# 如果是 multipart/form-data 编码方式,则当成表单数据解析,可用
# 于处理文件上传请求
elif self.content_type.startswith(b"multipart/form-data"):
    # 该正则用于取出数据分割符
    boundary_search_result = re.search(
        b"multipart/form-data; boundary=(.+)",
        self.content_type
    )
    if boundary_search_result:
        # 取出分割符
        boundary = boundary_search_result.group(1)
        if self._body:
            # 用分割符分割表单数据
            body_results = self.body.split(
                b'\r\n--' + boundary
            )
            if body_results:
                for body_result in body_results:
                    # 分割后的每一条数据都有头部和内容
                    # 头部和内容以 \r\n\r\n 分开
                    # 头部是字符串,描述该数据的信息
                    # 内容部分是二进制数据
                    split_index = body_result \
                        .find(b'\r\n\r\n')
                    if split_index != -1:
                        # 获取头部信息字符串
                        head = body_result[:split_index]
                        # 获取内容
                        content = body_result[split_index + 4:]
                        # 取出该字段的名称
                        name_result = re.search(
                          b'Content-Disposition: form-data; name="([^"]+)"',
                            head, re.M
                        )
                        if name_result:
                            name = name_result.group(1)
                            if name not in self._body_vars:
                                self._body_vars[name] = []
                            # 如果是文件,则取出该字段的文件名
                            file_name_result = re.search(
                                b' filename="([^"]+)"',
                                head, re.M
                            )
                            file_name = file_name_result \
                                .group(1) if \
                                file_name_result else None
```

```
                                if not file_name:
                                    # 如果不是文件,则值为普通字符串
                                    self._body_vars[name] \
                                        .append(content)
                                else:
                                    file_object = {
                                        'name': name,
                                        'filename': file_name,
                                        'content': content
                                    }
                                    content_type_result = \
                                        re.search(
                                            b'Content-Type: (.*)',
                                            head, re.M
                                        )
                                    if content_type_result:
                                        file_object[
                                            'content-type'
                                            ] = \content_type_result.group(1)
                                    # 如果是文件,则值为文件对象
                                    self._body_vars[name] \
                                        .append(file_object)
                            pass
                        else:
                            break
            else:
                Cms4pyLog.get_instance().info(
                    f"Request content-type is {self.content_type}, we do not parse"
                )
        else:
            Cms4pyLog.get_instance().warning("content-type is None")
        pass
    pass
```

为了能够使解析表单的功能生效,需要在 dynamic_handler.py 文件的 handle_dynamic_request 函数中执行该函数,将该代码写在执行 action 之前,代码如下:

```
await req._parse_form()
await action(req, res)
```

为了便于使用与表单解析相关的功能,还应该为 Request 类添加几个函数,代码如下:

```
@property
    def query_vars(self):
        """
获取所有 URL 中的参数对
        :return:
        """
        return self._query_vars
```

```
    def get_query_vars(self, key: Bytes) -> list:
        """
根据键名获取对应的所有的值
        :param key:
        :return:
        """
        return self.query_vars[key] if key in self.query_vars else None

    def get_query_var(self, key: Bytes, default_value=b'') -> Bytes:
        """
根据键名获取与之匹配的第一个值
        :param key:
        :param default_value:
        :return:
        """
        return self._get_first_value_of_array_map(
            self.query_vars, key
        ) or default_value

    @property
    def body_vars(self):
        """
获取通过 HTTP 消息体传来的参数
        :return:
        """
        return self._body_vars

    def get_body_vars(self, key: Bytes) -> list:
        """
根据键名获取对应的所有消息体参数
        :param key:
        :return:
        """
        return self._body_vars[key] if key in self._body_vars else None

    def get_body_var(self, key: Bytes, default_value=b'') -> Bytes:
        """
根据键名获取对应的第一个消息体参数
        :param key:
        :param default_value:
        :return:
        """
        return self._get_first_value_of_array_map(
            self.body_vars, key
        ) or default_value

    def get_var(self, key: Bytes, default_value=b'') -> Bytes:
```

```
        """
根据键名获取对应的第一个参数,该函数会自动从 URL 和消息体中读取参数
        :param key:
        :param default_value:
        :return:
        """
        if self.method == "GET":
            return self.get_query_var(
                key, default_value
            )
        elif self.method == 'POST':
            return self.get_body_var(
                key, default_value
            ) or self.get_query_var(
                key, default_value
            )
        else:
            return default_value

    def get_var_as_str(
            self, key: Bytes,
            default_value = '',
            charset = config.GLOBAL_CHARSET
    ) -> str:
        """
以指定的编码方式获取参数
        :param key:
        :param default_value:
        :param charset:
        :return:
        """
        var_Bytes = self.get_var(key)
        if var_Bytes:
            return unquote(var_Bytes.decode(charset))
        else:
            return default_value
```

在 Action 中使用该功能的示例代码如下：

```
async def login_handler(req, res):
    # 用于获取 URL 中的参数
    req.get_query_var(b"name")
    # 用于获取消息体中的参数
    req.get_body_var(b"name")
    # 自动从 URL 和消息体中获取参数
    req.get_var(b"name")
```

8.10 实现 Cookie 操作

Cookie 是浏览器存储用户信息的一种机制，并且会在每次请求时将该信息通过协议头发送到服务器。Cookie 可以存储多个键值对，以分号将它们隔开，其格式如：common_sid=xxx；SameSite=Lax。由于需要在浏览器与服务器之间频繁传输，所以 Cookie 被设计为不能用于存储较大的数据，否则将对服务器造成巨大压力。

在 Request 类中添加成员函数用于获取 Cookie 数据，代码如下：

```
"""
第 8 章/cms4py_first_generation/cms4py/http/request.py
"""

class Request:
    # 此处省略部分代码

    @property
    def cookie(self) -> Bytes:
        """
获取 Cookie 字符串
        :return:
        """
        if not self._cookie:
            # 从 header 里取出 Cookie
            self._cookie = self.get_header(b'cookie')
        return self._cookie

    def get_cookie(self, key: Bytes, default_value = None) -> Bytes:
        """
根据键名获取 Cookie 的值
        :param key:
        :param default_value:
        :return:
        """
        if not self._cookie_map:
            self._cookie_map = {}
            cookie = self.cookie
            if cookie:
                tokens = cookie.split(b"; ")
                for t in tokens:
                    kv = t.split(b"=")
                    if len(kv) == 2:
                        self._cookie_map[kv[0]] = kv[1]
        return self._cookie_map[key] if key in self._cookie_map else default_value
```

除了可以读取 Cookie 外，在 HTTP 协议中还定义了通过服务器端修改 Cookie 的方式，原理是在响应的协议头中写 Set-Cookie 字段。一个操作 Cookie 的示例代码如下：

```
"""
第 8 章/cms4py_first_generation/cms4py/http/response.py
"""

import config
from cms4py.http.request import Request


class Response:
    # 此处省略部分代码

    def add_set_cookie(
            self,
            name: Bytes,
            value: Bytes,
            max_age: int = 604800,
            path: Bytes = b'/'
    ):
        """
        添加设置 Cookie 的协议头
        :param name: Cookie 的名称
        :param value: Cookie 的值
        :param max_age: Cookie 的有效期
        :param path: Cookie 的指定路径
        :return:
        """

        self.add_header(
            b'set-cookie',
            "{}={}; max-age={}; path={}; SameSite=Lax".format(
                name.decode(config.GLOBAL_CHARSET),
                value.decode(config.GLOBAL_CHARSET),
                max_age,
                path.decode(config.GLOBAL_CHARSET)
            ).encode(
                config.GLOBAL_CHARSET
            )
        )
```

接下来写一个 Action 来测试 Cookie 操作功能，代码如下：

```
"""
第 8 章/cms4py_first_generation/app/controllers/cookie.py
"""


async def counter(req, res):
    count = int(req.get_cookie(b"count") or b'0')
    count += 1
```

```
    res.add_set_cookie(b'count', str(count).encode("utf-8"))
    await res.end(f"Count: {count}".encode("utf-8"))
```

通过浏览器访问页面 http://127.0.0.1:8000/cookie/counter，不断刷新该页面，可以看到 Cookie 中的 count 值不断自增，效果如图 8-10 所示。

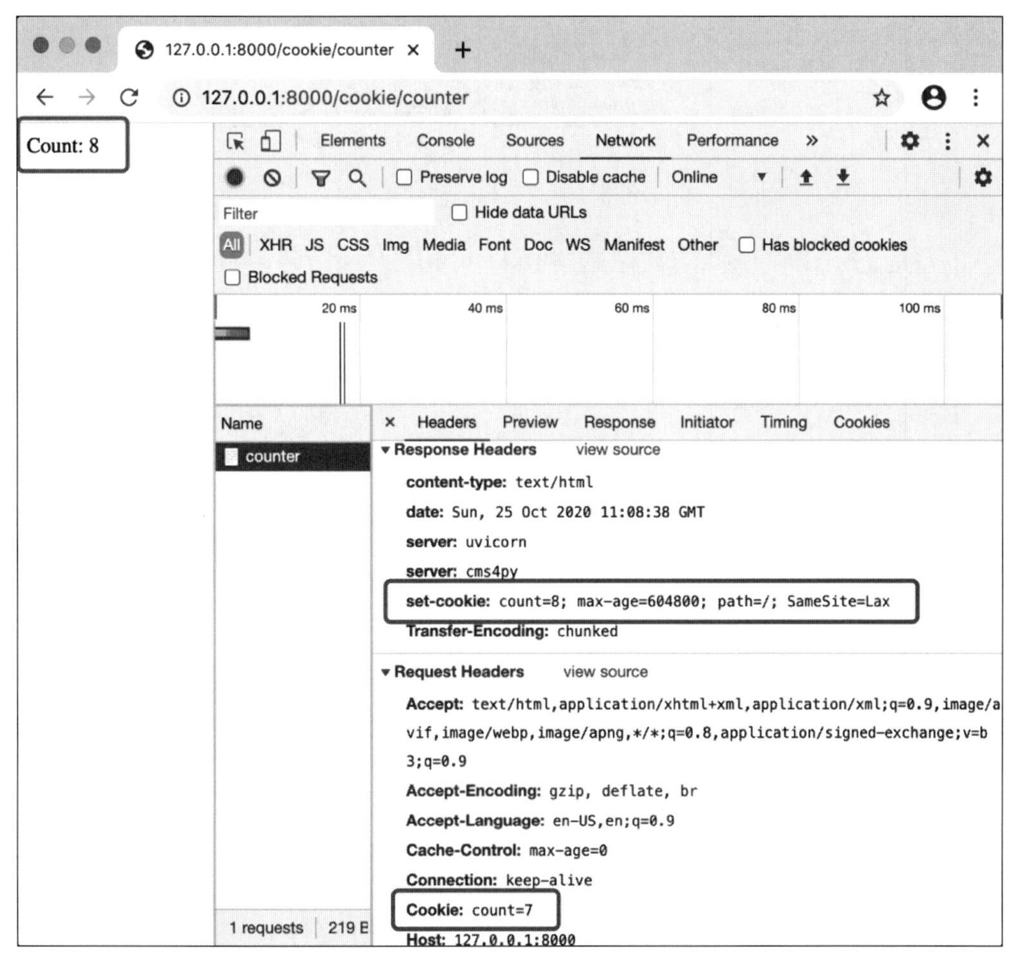

图 8-10 操作 Cookie 效果图

8.11 实现 Session 机制

Cookie 数据会随 HTTP 请求一起发送，除了数据量小之外，还容易被泄露，为了解决这些问题，需要实现一种机制，能够将用户信息存放在服务器端，与浏览器用户端对应，却无须在 HTTP 请求时传递，这就是 Session 机制。

Session 的实现原理流程如图 8-11 所示。

服务器端为每个浏览器用户生成一个对象用于存储用户数据，同时为访问该对象而生成唯一编码，并将该唯一编码 session_id 的键名存于 Cookie 中，当浏览器再次请求时，服务器根据该唯一编码可获取与之对应的对象。

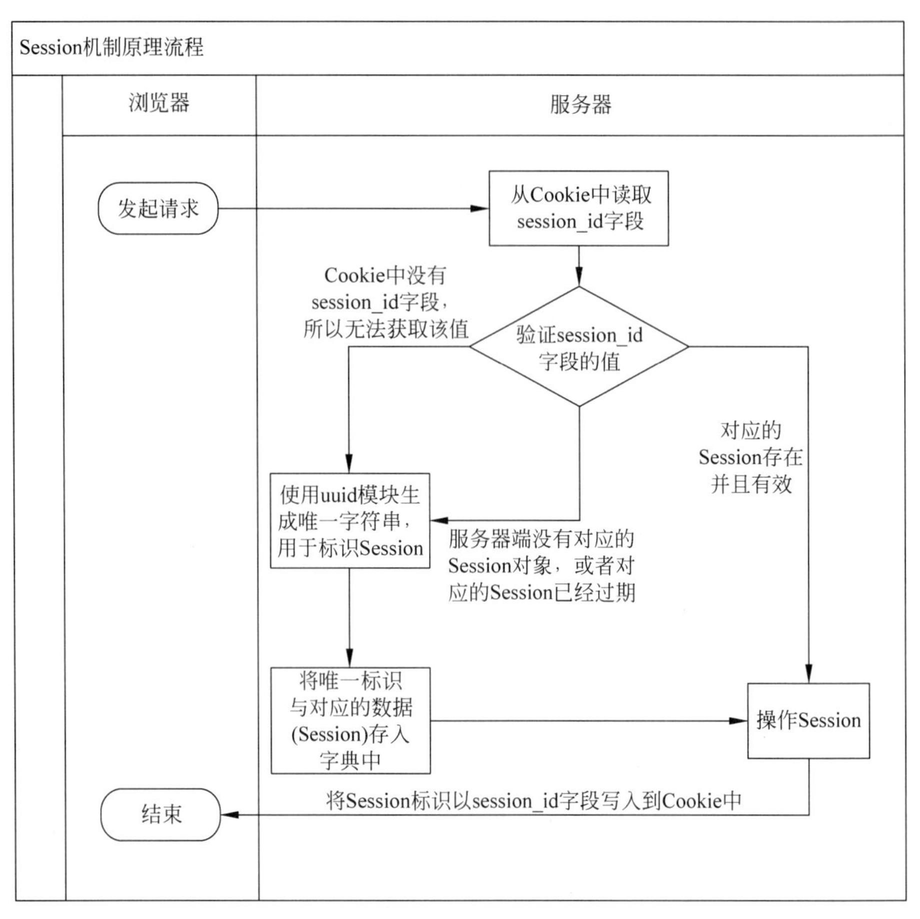

图 8-11　Session 机制原理流程图

为了实现该机制，首先实现核心缓存机制，代码如下：

```
"""
第 8 章/cms4py_first_generation/cms4py/cache/session_cache_manager.py
"""

from cms4py.cache.base_cache_manager import BaseCacheManager
from cms4py.cache.page_cache_manager import CachedDataWrapper
import datetime


class SessionCacheManager(BaseCacheManager):
    __instance = None

    @staticmethod
    def get_instance():
        if not SessionCacheManager.__instance:
            SessionCacheManager.__instance = SessionCacheManager()
        return SessionCacheManager.__instance

    async def will_reload(
```

```
            self,
            wrapper: CachedDataWrapper,
            key: str
    ) -> bool:
        # TODO 设计 Session 过期机制
        return False

    async def wrap_data(self, key) -> CachedDataWrapper:
        return CachedDataWrapper(
            {}, datetime.datetime.now().timestamp()
        )

    async def set_current_user(self, session_id: Bytes, user):
        """
设置当前登录的用户
        :param session_id:
        :param user:
        :return:
        """
        session = await self.get_data(session_id)
        session['current_user'] = user

    async def get_current_user(self, session_id: Bytes):
        """
获取当前登录的用户
        :param session_id:
        :return:
        """
        session = await self.get_data(session_id)
        return session['current_user'] \
            if 'current_user' in session else None
```

在 cache 模块中添加代码以使该类便于被外部访问，代码如下：

```
"""
第 8 章/cms4py_first_generation/cms4py/cache/__init__.py
"""

from cms4py.cache.session_cache_manager import SessionCacheManager
```

接下来在 Request 类中添加如下代码：

```
"""
第 8 章/cms4py_first_generation/cms4py/http/request.py
"""

import config
import re, uuid
from cms4py.utils.log import Cms4pyLog
from cms4py.utils import URL_helper
```

```
from URLlib.parse import unquote
from cms4py.cache import SessionCacheManager


class Request:
    """
此处省略部分代码
    """

    async def session(self):
        """
根据 Session ID 获得对应的 Session 数据
        :return:
        """
        return await SessionCacheManager.get_instance().get_data(
            self.session_id
        )

    async def get_session(self, key: str, default_value = None):
        """
从 Session 中根据键名获取对应的值
        :param key:
        :param default_value:
        :return:
        """
        session_dict = await SessionCacheManager.get_instance() \
            .get_data(self.session_id)
        return session_dict[key] \
            if key in session_dict else default_value

    async def set_session(self, key: str, value):
        """
将键值对写入 Session 中
        :param key:
        :param value:
        :return:
        """
        session_dict = await SessionCacheManager.get_instance() \
            .get_data(self.session_id)
        session_dict[key] = value
```

为了将 Session ID 存放在 Cookie 中，还应该在 Response 类中添加如下代码：

```
#将 Session ID 的值写在 Cookie 中
self.add_set_cookie(
    config.CMS4PY_SESSION_ID_KEY,
    self._request.session_id
)
```

使用该 Session 机制的代码如下：

```python
"""
第 8 章/cms4py_first_generation/app/controllers/session.py
"""


async def counter(req, res):
    count = int((await req.get_session("count")) or '0')
    count += 1
    await req.set_session("count", count)
    await res.end(f"Count: {count}".encode("utf-8"))
```

在 cms4py 中，Session 存储的默认实现是将数据存于内存（Python 字典）中，这种方式速度很快，但是不能持久化，这意味着服务器重启会导致所有用户数据失效。为了解决这个问题，可将数据直接存储在硬盘上，但是由于硬盘的读写速度远低于内存，所以可能会影响运行效率，这就要求实现一种机制先将数据缓存于内存中，然后另外开启一个单独的线程管理将数据同步到硬盘上的逻辑，以不影响网站程序的运行。

一种方案是将 Session 数据存放在数据库中。但由于每一次请求都会用到 Session 机制，如果 Session 与网站主要程序使用同一个数据库，那么 Session 机制频繁与数据库的交互会影响网站主要程序操作数据库的效率。所以一般情况下会使用一个单独的数据库用于存储 Session 数据，常见的选择是 MongoDB。

还有一种方案是将 Session 数据存放在高速缓存技术中，如 Redis 或者 Memcached，也可以将数据持久化存放在硬盘上。由于高速缓存技术的实现逻辑比数据库更简单，所以运行速度更快，是用于存储 Session 数据的不错方案。

8.12 实现多语言支持

根据 config.py 文件中的配置，在 app 目录下创建一个 languages 目录，用于存放语言文件。在 cms4py 中语言文件采用 JSON 格式，并且采用文件名对应语言名的方式，如图 8-12 所示。

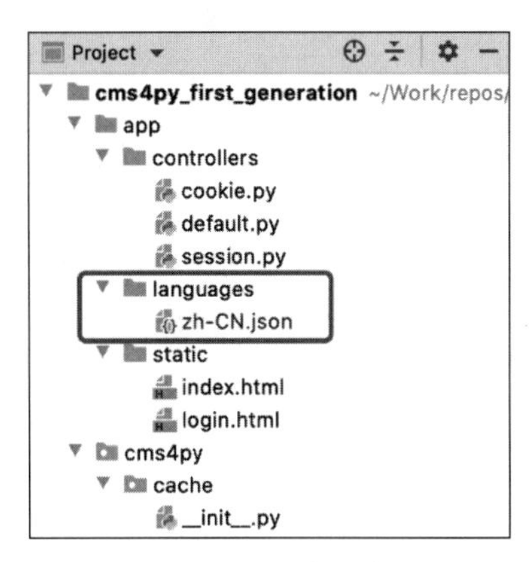

图 8-12 语言文件所在目录

在网站运行过程中，语言文件一般不会发生改变，如果每次请求都读取语言文件并解析，将会造成资源浪费，所以需要文件缓存机制，实现代码如下：

```python
"""
第 8 章/cms4py_first_generation/cms4py/cache/file_cache_manager.py
"""
```

```
from cms4py.cache.base_cache_manager import BaseCacheManager
from cms4py.cache.base_cache_manager import CachedDataWrapper
from cms4py.utils import aiofile


class FileCacheManager(BaseCacheManager):
    __instance = None

    @staticmethod
    def get_instance() -> "FileCacheManager":
        if not FileCacheManager.__instance:
            FileCacheManager.__instance = FileCacheManager()
        return FileCacheManager.__instance

    def __init__(self) -> None:
        super().__init__()

    async def wrap_data(self, key) -> CachedDataWrapper:
        wrapper = None
        if await aiofile.exists(key) and await aiofile.isfile(key):
            content = await aiofile.read_file(key)
            # 将文件与对应的修改时间包装成缓存数据包装器
            wrapper = CachedDataWrapper(content, await aiofile.getmtime(key))
        return wrapper

    async def will_reload(self, wrapper: CachedDataWrapper, key: str) -> bool:
        """
如果文件已修改,则重新加载文件
        :param wrapper: 已缓存的文件数据包装器
        :param key: 文件路径
        :return:
        """
        return await aiofile.exists(
            key
        ) and await aiofile.isfile(
            key
        ) and await aiofile.getmtime(
            key
        ) != wrapper.timestamp
```

语言文件设计为JSON格式的键值对列表,内容如下:

```
{
  "Hello": "你好",
  "Home": "主页"
}
```

在程序中需要直接将该JSON格式的内容转换成Python对象以便使用,实现代码如下:

```
"""
第 8 章/cms4py_first_generation/cms4py/cache/json_file_cache_manager.py
"""

from cms4py.cache.file_cache_manager import FileCacheManager
from cms4py.cache.base_cache_manager import CachedDataWrapper
from cms4py.utils.log import Cms4pyLog
import json

class JsonFileCacheManager(FileCacheManager):
    __instance = None

    @staticmethod
    def get_instance() -> "JsonFileCacheManager":
        if not JsonFileCacheManager.__instance:
            JsonFileCacheManager.__instance = JsonFileCacheManager()
        return JsonFileCacheManager.__instance

    async def wrap_data(self, key) -> CachedDataWrapper:
        wrapper = await super().wrap_data(key)
        if wrapper and wrapper.data:
            #对数据进行重新包装,将文件内容转换成 json 对象
            wrapper.data = json.loads(wrapper.data)
            Cms4pyLog.get_instance().debug(f"Load json file {key}")
        return wrapper
```

接下来实现一个翻译工具,用于加载和翻译文字,代码如下:

```
"""
第 8 章/cms4py_first_generation/cms4py/utils/translator.py
"""

import config, os
from cms4py.utils import aiofile
from cms4py.cache.json_file_cache_manager import JsonFileCacheManager


async def get_language_dict(language):
    """
根据语言名将对应的 JSON 文件读成 Python 对象
    :param language: 语言名,如:zh-CN、en-US 等
    :return: 如果文件不存在,则返回 None
    """
    language_dict = None
    language_file = os.path.join(
        config.LANGUAGES_ROOT, f"{language}.json"
    )
    if await aiofile.exists(
            language_file
```

```
        ) and await aiofile.isfile(language_file):
            language_dict = await JsonFileCacheManager \
                .get_instance() \
                .get_data(language_file)
        return language_dict


async def translate(words: str, target_language: str) -> str:
    """
将指定的句子翻译成目标语言
    :param words:
    :param target_language:
    :return:
    """
    result = words
    #加载语言表
    language_dict = await get_language_dict(target_language)
    if language_dict and words in language_dict:
        #获取语言对应的翻译
        result = language_dict[words]
    return result
```

该工具的实现基于异步IO,但是目前在模板中使用异步IO代码有些困难,所以为了兼容后续集成的模板渲染功能,这里需要对语言文件进行预先加载。在Response类中添加如下代码:

```
"""
第8章/cms4py_first_generation/cms4py/http/response.py
"""

import config
from cms4py.http.request import Request
from cms4py.utils import translator


class Response:
    #此处省略部分代码

    async def _load_language_dict(self):
        """
加载语言表
        :return:
        """
        if not self._language_dict:
            self._language_dict = await translator \
                .get_language_dict(self._request.language)
```

```
    def translate(self, words):
        """
在不使用异步 IO 的情况下进行翻译
        :param words:
        :return:
        """
        if self._language_dict and words in self._language_dict:
            words = self._language_dict[words]
        return words
```

然后在 handle_dynamic_request 函数中添加代码用于预加载语言表，代码如下：

```
await res._load_language_dict()
```

此时 dynamic_handler.py 文件代码如下：

```
"""
第 8 章/cms4py_first_generation/cms4py/handlers/dynamic_handler.py
"""

import inspect

import config
from cms4py import http
from cms4py.cache import ModulesCacheManager


async def handle_dynamic_request(scope, receive, send) -> bool:
    data_sent = False
    request_path: str = scope['path']

    # 将请求路径分拆为 Controller 和 Action,例如:/user/list_all 请求
    # 对应的 Controller 是 user,而 Action 是 list_all
    tokens = request_path.split("/")
    tokens_len = len(tokens)
    # 指定默认的控制器名
    controller_name = config.DEFAULT_CONTROLLER
    if tokens_len >= 2:
        controller_name = tokens[1] or config.DEFAULT_CONTROLLER
    # 指定默认的函数名
    action_name = config.DEFAULT_ACTION
    if tokens_len >= 3:
        action_name = tokens[2] or config.DEFAULT_ACTION

    controller_object = await ModulesCacheManager.get_instance().get_data(
        f"{config.APP_DIR_NAME}.{config.CONTROLLERS_DIR_NAME}.{controller_name}"
    )

    if controller_object:
        # 根据 action_name 获取指定的成员
        action = getattr(controller_object, action_name, None)
```

```
        if action:
            #构造 HTTP 请求对象,便于后续操作
            req = http.Request(scope, receive)
            #构造 HTTP 响应对象,便于后续操作
            res = http.Response(req, send)
            req._controller = controller_name
            req._action = action_name

            #将 action 后的路径参数记录在 req.args 中
            req._args = tokens[3:] if len(tokens) > 3 else []
            #如果 action 是类定义,则先将类实例化再执行
            if inspect.isclass(action):
                await req._parse_form()
                #预加载语言表
                await res._load_language_dict()
                await action()(req, res)
            #否则把 action 当作函数对待并直接执行
            else:
                await req._parse_form()
                #预加载语言表
                await res._load_language_dict()
                await action(req, res)
            data_sent = res.body_sent
    return data_sent
```

接下来编写一个 Action,用于测试翻译效果,代码如下:

```
"""
第 8 章/cms4py_first_generation/app/controllers/lang.py
"""


async def hello(req, res):
    words = res.translate("Hello")
    await res.end(words.encode("utf-8"))
```

分别使用英文浏览器和中文浏览器,访问效果如图 8-13 所示。

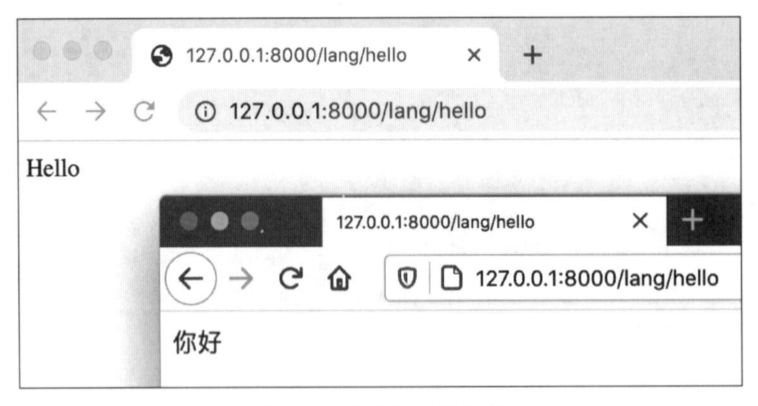

图 8-13　中英文翻译效果

8.13 集成模板渲染功能

模板渲染功能可以大大提高网站的开发速度，是 cms4py 中不可或缺的功能。模板渲染功能的编写难度比较大，涉及语法分析等技术，而且工作量也比较大，所以直接集成现有的成熟模板渲染技术 Jinja2。

接下来通过命令 pip install jinja2 安装该模板引擎，为了方便在 cms4py 框架中使用 Jinja2，需要先设计一个类，以便对 Jinja2 模板引擎进行管理，代码如下：

```
"""
第 8 章/cms4py_first_generation/cms4py/template_engine.py
"""

from jinja2 import FileSystemLoader, Environment

import config
import asyncio


class TemplateEngine:
    __instance = None

    @staticmethod
    def get_instance():
        if not TemplateEngine.__instance:
            TemplateEngine.__instance = TemplateEngine()
        return TemplateEngine.__instance

    def __init__(self) -> None:
        super().__init__()
        #配置 jinja2 环境,用于渲染模板
        self._jinja2_env = Environment(
            loader = FileSystemLoader(config.VIEWS_ROOT)
        )

    async def render(self, view, **kwargs) -> Bytes:
        """
将模板渲染成字节数组
        :param view:
        :param kwargs:
        :return:
        """
        return (
            await self.render_to_string(view, **kwargs)
        ).encode(config.GLOBAL_CHARSET)

    async def render_to_string(self, view, **kwargs) -> str:
```

```
        """
将模板渲染成字符串
        :param view:
        :param kwargs:
        :return:
        """
        rloop = asyncio.get_running_loop()
        tpl = await rloop.run_in_executor(
            None, self._jinja2_env.get_template, view
        )
        return await rloop.run_in_executor(None, tpl.render, kwargs)
```

接下来在 Response 类中添加成员函数用于在 Action 中调用，代码如下：

```
"""
第 8 章/cms4py_first_generation/cms4py/http/response.py
"""

import config
from cms4py.http.request import Request
from cms4py.utils import translator
from cms4py.template_engine import TemplateEngine


class Response:
#此处省略部分代码

    async def render_string(self, view: str, **kwargs) -> Bytes:
        """
渲染一段字符串
        :param view: 将被渲染的模板字符串
        :param kwargs: 向模板渲染过程传参数
        :return: 被渲染之后的数据
        """

        #该参数用于在模板渲染过程中访问配置信息
        kwargs['config'] = config
        #该参数用于在模板渲染过程中访问 Response 对象
        kwargs['response'] = self
        #该参数用于在模板渲染过程中访问 Request 对象
        kwargs['request'] = self._request
        #该参数用于在模板渲染过程中访问翻译工具对象
        kwargs["_"] = self.translate
        #该参数用于在模板渲染过程中访问翻译工具对象
        kwargs["T"] = self.translate
        #该参数用于在模板渲染过程中访问 Session
        kwargs['session'] = await self._request.session()
        data = await TemplateEngine.get_instance().render(
            view, **kwargs
```

```
        )
        return data

    async def render(self, view: str, **kwargs):
        """
渲染模板文件,并返回浏览器端
        :param view:
        :param kwargs:
        :return:
        """
        if self._body_sent:
            return
        await self.end(
            await self.render_string(view, **kwargs)
        )
```

接下来写一个 Action,用于测试模板渲染功能,代码如下:

```
"""
第 8 章/cms4py_first_generation/app/controllers/home.py
"""

async def index(req, res):
    await res.render("home/index.html")
```

与之对应的模板文件内容如下:

```
<!-- 第 8 章/cms4py_first_generation/app/views/home/index.html -->
<!DOCTYPE html>
<html lang="en">
<head>
<meta charset="UTF-8">
<title>Title</title>
</head>
<body>
模板文件内容
</body>
</html>
```

渲染结果如图 8-14 所示。

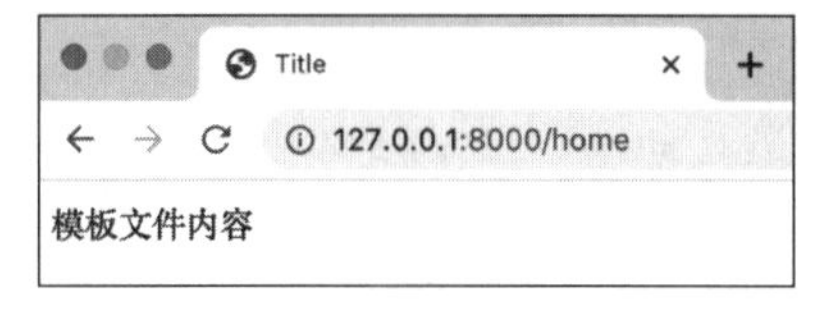

图 8-14 集成模板功能测试效果

8.14 实现页面重定向

重定向状态有3种，分别为301 Moved Permanently、302 Found和303 See Other。

301 Moved Permanently表示永久重定向，说明请求的资源已经被永久移动到了新的地址。当浏览器在访问该页面被重定向到新页面后，当再次请求该页面时，浏览器不再向服务器发送该页面的请求，而是直接请求新页面。搜索引擎在抓取到该页面被永久重定向后，会将原收录页面替换为新页面，所以在进行网站页面迁移时经常会使用301跳转，从而保证网站在页面迁移过程中不被搜索引擎降低权重。

302 Found表示临时重定向，说明请求的资源已经被临时移动到了新的地址，搜索引擎不更新原收录页面。

303 See Other通常用于呈现上传文件的返回结果页面以替代指向刚上传成功的文件链接。

接下来将状态码301和302封装为一个函数，以便后续调用，该函数写在Response类中，代码如下：

```
"""
第8章/cms4py_first_generation/cms4py/http/response.py
"""

import config
from cms4py.http.request import Request
from cms4py.utils import translator
from cms4py.template_engine import TemplateEngine


class Response:
    # 此处省略部分代码

    async def redirect(self, target: str, primary = False):
        """
页面重定向
        :param target: 目标地址
        :param primary: 是否为永久重定向
        :return:
        """
        URL = target.encode(config.GLOBAL_CHARSET)
        status = 302 if not primary else 301
        self.add_header(b'location', URL)
        await self.send_header(status)
        await self.end(
            b"<html lang = \"en\">"
            b"    <head>"
            b"        <meta charset = \"UTF-8\">"
            b"        <title>Redirecting</title>"
```

```
            b"    </head>"
            b"    <body>"
            b"        Redirecting to <a href='" + URL + b"'>" + URL + b"</a>" +
            b" </body>"
            b"</html>")
        pass
```

8.15 集成 pyDAL

pyDAL 来自 web2py，最初集成在 web2py 中，后被作者分离出来，是一个非常优秀的 Python 语言数据库抽象层，可支持绝大多数主流数据库。

在集成 pyDAL 到 cms4py 中之前，首先来学习如何使用 pyDAL 操作数据库。

在使用命令 pip install pydal 安装该依赖项之后，可使用 DAL 函数连接数据库，之后便可以基于该 DAL 对象定义与操作数据库，绝大多数工作已自动完成，用法极为简单，代码如下：

```
"""
第 8 章/pydal_example/app.py
"""

from pydal import DAL, Field

if __name__ == '__main__':
    # 连接数据库，默认使用 SQLite 数据库引擎
    # 生成 dummy.db 文件，如果要修改已保存的文件 data.db
    # 则写法为 DAL("sqlite://data.db")
    db = DAL()
    # 定义名为 student 的表，pyDAL 会自动生成
    # 自增 id 列
    db.define_table(
        'student',
        Field('name'),
        Field('age')
    )
    # 将数据保存
    db.commit()
    # 向 student 表中插入一条数据
    db.student.insert(
        name = "小云",
        age = "20"
    )
    # 从 student 表中查询出所有数据
    all_students = db(db.student.id > 0).select()

    print(all_students)
    pass
```

执行这段代码将自动创建数据库，如图 8-15 所示。

其中连接数据库的字符串格式为数据库类型://用户名:密码@主机/库名。例如，以 root 用户和 rootpw 密码连接本机(127.0.0.1)的 MySQL 数据库中的 mydb 库，则写法为 MySQL://root:rootpw@127.0.0.1/mydb。

图 8-15 pyDAL 自动创建数据库

接下来用 docker-compose 创建一个 MariaDB 环境，docker-compose.yml 文件内容如下：

```
#第 8 章/pydal_example/docker-compose.yml

version: '3.1'

services:

  db:
    #数据库镜像使用 mariadb
    image: mariadb
    restart: always
    environment:
      #配置 root 用户，密码为 rootpw
      MYSQL_ROOT_PASSWORD: rootpw
    ports:
      #将端口映射到本机
      - 3306:3306

  adminer:
    #adminer 是一个数据管理系统
    image: adminer
    restart: always
    ports:
      #将 adminer 的端口映射到本机，便于访问
      - 8080:8080
```

使用命令 docker-compose up -d 启动这两个 Docker 服务，如图 8-16 所示。

```
Terminal:  Local ×  +
(venv) yunp.top@yunptops-MBP pydal_example % docker-compose up -d
Creating pydal_example_db_1      ... done
Creating pydal_example_adminer_1 ... done
(venv) yunp.top@yunptops-MBP pydal_example %
```

图 8-16 启动数据库服务

通过网址 http://127.0.0.1:8080 访问 Adminer，选择主机为 db，输入用户名为 root，密码为 rootpw，登录数据库管理系统，如图 8-17 所示。

登录后创建 mydb 库，如图 8-18 所示。

不用创建表，pyDAL 会自动创建缺失的表。连接 MariaDB 数据库的代码如下：

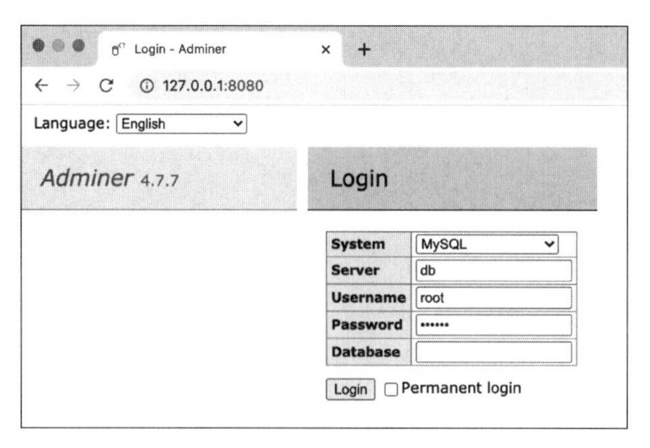

图 8-17　登录 Adminer

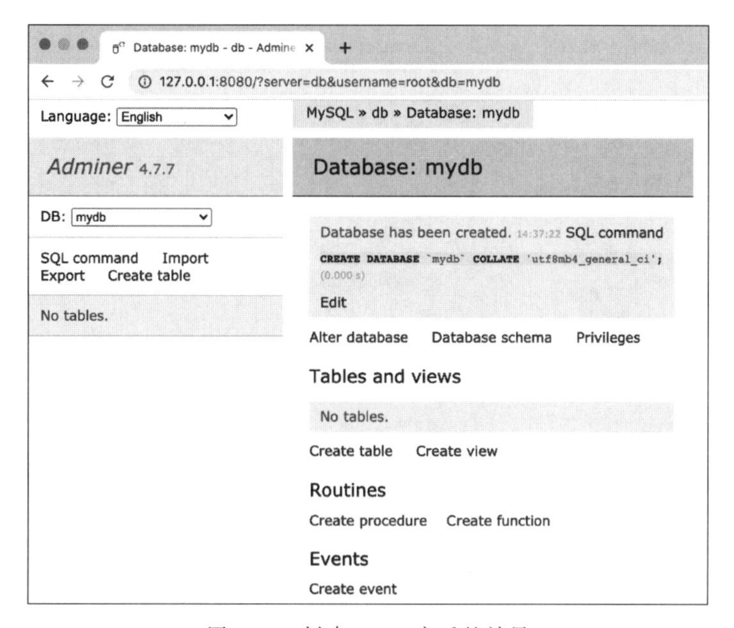

图 8-18　创建 mydb 库后的效果

```
"""
第 8 章/pydal_example/app.py
"""

from pydal import DAL, Field

if __name__ == '__main__':
    db = DAL("MySQL://root:rootpw@127.0.0.1/mydb")
    db.define_table(
        'student',
        Field('name'),
        Field('age')
    )
    db.student.insert(
```

```
        name = "小云",
        age = "20"
    )
    db.commit()
    all_students = db(db.student.id > 0).select()
    print(all_students)
    pass
```

由于连接的数据库是 MariaDB，所以需要选择数据库驱动 pyMySQL，使用命令 pip install pyMySQL 安装该驱动，然后运行该代码，相关表及数据会自动生成，如图 8-19 所示。

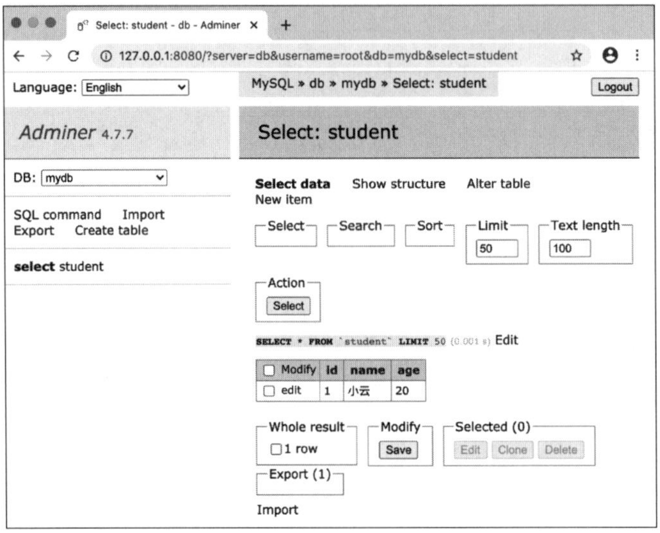

图 8-19 pyDAL 操作 MariaDB 后的效果

虽然 pyDAL 功能非常强大并且用起来得心应手，但是遗憾的是由于它是基于阻塞型 IO 的，以及软件作者目前还没有开发支持异步 IO 的计划，所以无法直接用在异步 IO 编程中。

幸运的是，pyDAL 支持将 pyDAL 表达式转换成 SQL 语句，这就免去了动手写数据库抽象层(DAL)的麻烦。利用 pyDAL 强大的 SQL 语句生成功能搭配 aioMySQL 驱动，可以将 pyDAL 完美地集成到 cms4py 中。

pyDAL 生成 SQL 语句的用法极为简单，只需在相应的指令函数前添加下画线便可生成 SQL 语句而不执行，示例代码如下：

```
"""
第 8 章/pydal_example/gen_sql.py
"""
from pydal import DAL, Field

if __name__ == '__main__':
    db = DAL("MySQL://root:rootpw@127.0.0.1/mydb")
    db.define_table(
        'student',
        Field('name'),
```

```
        Field('age')
    )
    #在相应的语句前添加下画线用于生成 SQL 语句而不执行
    sql = db.student._insert(
        name = "小云", age = "20"
    )
    print(sql)

    sql = db(db.student.id == 2)._update(
        name = "小明", age = "10"
    )
    print(sql)

    sql = db(db.student.id > 0)._select()
    print(sql)

    sql = db(db.student.id == 1)._delete()
    print(sql)
```

代码运行输出效果如图 8-20 所示。

```
gen_sql ×
/Users/yunp.top/Work/repos/github/plter/PythonAIOProgramming/第8章/pydal_example/venv/bin/python /Users
INSERT INTO `student`(`age`,`name`) VALUES ('20','小云');
UPDATE `student` SET `age`='10',`name`='小明' WHERE (`student`.`id` = 2);
SELECT `student`.`id`, `student`.`name`, `student`.`age` FROM `student` WHERE (`student`.`id` > 0);
DELETE `student` FROM `student` WHERE (`student`.`id` = 1);
```

图 8-20　pyDAL 生成 SQL 效果

由于仅仅使用了 pyDAL 的生成 SQL 功能，所以如果使用另一套机制（aioMySQL）去操作数据库，则须根据 pyDAL 的表定义手动建立对应的数据库结构（参考第 4 章 aioMySQL）。

原理已知，但是实现起来工作量较大，仍然需要一定的耐心。

项目文件结构如图 8-21 所示。

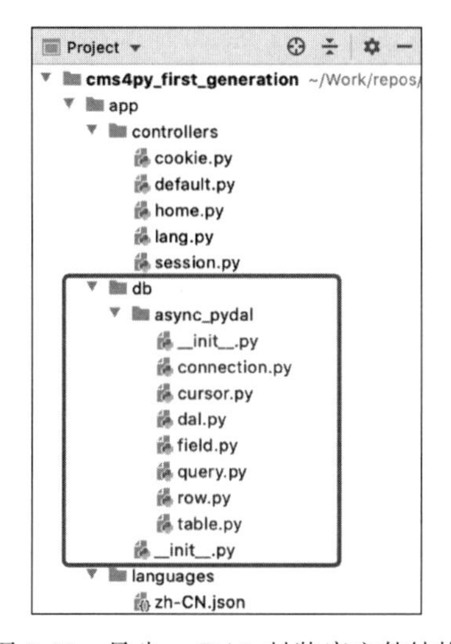

图 8-21　异步 pyDAL 封装库文件结构

文件 dal.py 代码如下：

```
"""
第 8 章/cms4py_first_generation/app/db/async_pydal/dal.py
"""

from pydal import DAL as PY_DAL

import aioMySQL
from aioMySQL.pool import Pool
from .connection import PyDALConnection


class AsyncDAL:
    """
该类将 pyDAL 封装为支持异步 IO 的模式,数据库驱动采用 aioMySQL
    @see https://aioMySQL.readthedocs.io/en/latest/pool.html
    """

    def __init__(self) -> None:
        super().__init__()
        self._pydal = None
        self._aioMySQL_conn_pool = None

    @staticmethod
    async def create(
            host, user, password, db, port = 3306,
            pool_max_size = 10, echo = True
    ) -> "AsyncDAL":
        """
该工厂方法用于支持以异步 IO 编程的方式创建 AsyncDAL 实例
        :param host: 数据库服务器主机
        :param user: 数据库登录用户名
        :param password: 数据库登录用户名对应的密码
        :param db: 数据库名
        :param port: 数据库服务器所在的端口号
        :param pool_max_size: 数据库连接池支持的最大连接数
        :param echo: 是否输出 SQL 语言,该功能可用于开发调试阶段
        :return:
        """
        return await AsyncDAL().async_init(
            host, user, password, db, port,
            pool_max_size, echo
        )

    async def async_init(
            self, host, user, password, db, port,
            pool_max_size, echo
    ) -> "AsyncDAL":
        """
```

```
以异步方式初始化该实例，由于内置的 __init__ 函数无法通过
异步 IO 的方式调用，所以写了个单独的支持以异步 IO 的方式调用的
函数进行初始化
        """

        # 配置 pyDAL 连接，在不执行数据库指令的情况下 pyDAL 并不连接数据库
        self._pydal = PY_DAL(
            uri = f"MySQL://{user}:{password}@{host}:{port}/{db}",
            migrate = False, migrate_enabled = False, bigint_id = True
        )

        # 配置 aioMySQL 连接
        self._aioMySQL_conn_pool = await aioMySQL.create_pool(
            0, pool_max_size, echo, host = host, port = port, user = user,
            password = password, db = db
        )
        return self

    @property
    def aioMySQL_conn_pool(self) -> Pool:
        """
获取 aioMySQL 连接池对象
        :return:
        """
        return self._aioMySQL_conn_pool

    def close(self):
        """
关闭连接池
        :return:
        """
        self.aioMySQL_conn_pool.close()

    async def wait_closed(self):
        """
等待连接池关闭
        :return:
        """
        await self.aioMySQL_conn_pool.wait_closed()

    def terminate(self):
        """
中止连接池
        :return:
        """
        self.aioMySQL_conn_pool.terminate()

    async def acquire(self) -> PyDALConnection:
        """
通过连接池获得一个连接上下文对象，并将其封装
```

```
到 PyDALConnection 中
        :return:
        """
        return PyDALConnection(
            self._pydal,
            await self.aioMySQL_conn_pool.acquire()
        )

    async def release(self, conn: PyDALConnection):
        """
释放一个连接对象
        :param conn:
        :return:
        """
        await self.aioMySQL_conn_pool.release(
            conn.aioMySQL_conn
        )

    def define_table(self, tablename, *fields, **kwargs):
        """
定义数据表
        """
        self._pydal.define_table(tablename, *fields, **kwargs)
```

文件 connection.py 代码如下：

```
"""
第 8 章/cms4py_first_generation/app/db/async_pydal/connection.py
"""

from aioMySQL.connection import Connection
from .cursor import PyDALCursor
from .table import Table


class PyDALConnection:
    """
数据库连接对象
    """

    def __init__(self, pydal, aioMySQL_conn: Connection) -> None:
        super().__init__()
        self._pydal = pydal
        self._aioMySQL_conn = aioMySQL_conn

    @property
    def aioMySQL_conn(self) -> Connection:
        return self._aioMySQL_conn
```

```
    async def commit(self):
        """
提交数据
        :return:
        """
        await self.aioMySQL_conn.commit()

    def close(self):
        """
关闭该连接
        :return:
        """
        self.aioMySQL_conn.close()

    async def autocommit(self, mode: bool):
        """
设置该连接为自动提交数据
        :param mode:
        :return:
        """
        await self.aioMySQL_conn.autocommit(mode)

    def autocommit_mode(self):
        """
用于确定当前是否为自动提交状态
        :return:
        """
        return self.aioMySQL_conn.autocommit_mode

    async def cursor(self) -> PyDALCursor:
        """
        Cursor 对象用于操作数据库
        :return:
        """
        return PyDALCursor(
            self._pydal,
            self,
            await self.aioMySQL_conn.cursor()
        )
```

文件 cursor.py 代码如下：

```
"""
第 8 章/cms4py_first_generation/app/db/async_pydal/cursor.py
"""

from aioMySQL.cursors import Cursor
from .table import Table
from .query import AsyncQuery
```

```
class PyDALCursor:
    def __init__(self, pydal, pydal_connection, aioMySQL_cursor: Cursor) -> None:
        super().__init__()
        self._aioMySQL_cursor = aioMySQL_cursor
        self._pydal = pydal
        self._pydal_connection = pydal_connection
        self._table_map = {}

    @property
    def aioMySQL_cursor(self) -> Cursor:
        return self._aioMySQL_cursor

    @property
    def pydal_connection(self):
        return self._pydal_connection

    @property
    def description(self):
        """
获取描述信息
        :return:
        """
        return self._aioMySQL_cursor.description

    async def execute(self, sql, args = None):
        """
以异步方式执行一条 SQL 语句
        :param sql: SQL 语句
        :param args: 与 SQL 语句对应的参数
        :return:
        """
        await self.aioMySQL_cursor.execute(sql, args)

    async def executemany(self, sql, args = None):
        await self.aioMySQL_cursor.executemany(sql, args)

    async def fetchall(self):
        """
获取执行 SQL 后得到的所有结果数据
        :return:
        """
        return await self.aioMySQL_cursor.fetchall()

    async def fetchone(self):
        """
获取一条数据
        :return:
        """
        return await self.aioMySQL_cursor.fetchone()
```

```
    async def fetchmany(self, size = None):
        return await self.aioMySQL_cursor.fetchmany(size)

    async def close(self):
        """
关闭该 Cursor 对象
        :return:
        """
        await self.aioMySQL_cursor.close()

    @property
    def lastrowid(self):
        return self.aioMySQL_cursor.lastrowid

    @property
    def rowcount(self) -> int:
        """
获取 SQL 语句影响的行数
        :return:
        """
        return self.aioMySQL_cursor.rowcount

    def __getitem__(self, item):
        return self.__getattr__(item)

    def __getattr__(self, item):
        """
重载获取成员的运算符,用于封装对 pyDAL 的操作
        :param item:
        :return:
        """
        pydal_table = self._pydal.__getattr__(item)
        if pydal_table in self._table_map:
            async_table = self._table_map[pydal_table]
        else:
            async_table = Table(self._pydal, self, pydal_table)
            self._table_map[pydal_table] = async_table
        return async_table

    def __call__(self, * args) -> AsyncQuery:
        """
在 pyDAL 中,db()函数会生成语句,重载该函数以便调用操作
用于封装对 pyDAL 对象的调用
        :param args:
        :return:
        """
        return AsyncQuery(self, self._pydal( * args))
```

文件 query.py 代码如下：

```
"""
第 8 章/cms4py_first_generation/app/db/async_pydal/query.py
"""

from . import row


class AsyncQuery:
    """
该对象用于封装 pyDAL 表达式
    """

    def __init__(self, pydal_cursor, dal_query) -> None:
        super().__init__()
        self._pydal_cursor = pydal_cursor
        self._dal_query = dal_query

    def _select(self, *args, **kwargs):
        return self._dal_query._select(*args, **kwargs)

    async def select(self, *args, **kwargs):
        sql: str = self._select(*args, **kwargs)
        sql = sql.replace("\\", "")
        #print(sql)
        await self._pydal_cursor.execute(sql)
        result = await self._pydal_cursor.fetchall()
        return row.Rows(self._pydal_cursor.description, result)

    def _update(self, **kwargs):
        return self._dal_query._update(**kwargs)

    async def update(self, **kwargs) -> int:
        sql = self._update(**kwargs)
        #print(sql)
        await self._pydal_cursor.execute(sql)
        return self._pydal_cursor.rowcount

    def _delete(self):
        return self._dal_query._delete()

    async def delete(self):
        sql = self._delete()
        #print(sql)
        await self._pydal_cursor.execute(sql)
        return self._pydal_cursor.rowcount

    def _count(self):
```

```
        return self._dal_query._count()

    async def count(self) -> int:
        sql = self._count()
        sql = sql.replace("\\", "")
        #print(sql)
        await self._pydal_cursor.execute(sql)
        result = await self._pydal_cursor.fetchone()
        return result[0]

    async def isempty(self):
        return await self.count() == 0
```

文件 table.py 代码如下：

```
"""
第 8 章/cms4py_first_generation/app/db/async_pydal/table.py
"""

from .field import Field


class Table:
    """
该对象用于封装 pyDAL 表达式
    """

    def __init__(self, pydal, pydal_cursor, pydal_table) -> None:
        """
        :param pydal: The pyDAL object
        :param pydal_connection: The PyDALConnection instance
        :param pydal_table:
        """
        super().__init__()
        self._pydal_table = pydal_table
        self._fields_map = {}
        self._pydal = pydal
        self._pydal_cursor = pydal_cursor

    @property
    def pydal_cursor(self):
        return self._pydal_cursor

    async def insert(self, **kwargs) -> int:
        sql = self._pydal_table._insert(**kwargs)
        await self.pydal_cursor.execute(sql)
        return self.pydal_cursor.rowcount
```

```
    async def update(self, query, **kwargs):
        return await self.pydal_cursor(query).update(**kwargs)

    async def delete(self, query):
        return await self.pydal_cursor(query).delete()

    async def update_or_insert(self, query, **kwargs):
        if await self.pydal_cursor(query).isempty():
            return await self.insert(**kwargs)
        else:
            return await self.update(query, **kwargs)

    def __getitem__(self, item):
        return self.__getattr__(item)

    def __getattr__(self, item):
        pydal_field = self._pydal_table[item]
        if pydal_field in self._fields_map:
            async_field = self._fields_map[pydal_field]
        else:
            async_field = Field(self._pydal, self._pydal_cursor, pydal_field)
            self._fields_map[pydal_field] = async_field
        return async_field

    @property
    def ALL(self):
        return self._pydal_table.ALL

    async def by_id(self, record_id):
        return (await self._pydal_cursor(self.id == record_id).select()).first()
```

文件 field.py 代码如下：

```
"""
第8章/cms4py_first_generation/app/db/async_pydal/field.py
"""


class Field:
    """
    该对象用于封装 pyDAL 表达式
    """

    def __init__(self, pydal, pydal_cursor, pydal_field) -> None:
        super().__init__()
        self._pydal_field = pydal_field
        self._pydal = pydal
        self._pydal_cursor = pydal_cursor
```

```
    def __eq__(self, other):
        if not isinstance(other, Field):
            return self._pydal_field.__eq__(other)
        else:
            return self._pydal_field.__eq__(other._pydal_field)

    def __ne__(self, other):
        return self._pydal_field.__ne__(other)

    def __gt__(self, other):
        return self._pydal_field.__gt__(other)

    def __ge__(self, other):
        return self._pydal_field.__ge__(other)

    def __lt__(self, other):
        return self._pydal_field.__lt__(other)

    def __le__(self, other):
        return self._pydal_field.__le__(other)

    def __invert__(self):
        return self._pydal_field.__invert__()

    def __str__(self):
        return self._pydal_field.__str__()

    def __repr__(self):
        return self._pydal_field.__repr__()

    def like(self, expression, case_sensitive = True, escape = None):
        query = self._pydal_field.like(expression, case_sensitive, escape)
        return query

    @property
    def requires(self):
        return self._pydal_field.requires

    @property
    def type(self):
        return self._pydal_field.type

    def with_alias(self, alias):
        return self._pydal_field.with_alias(alias)
```

文件 row.py 代码如下：

```
"""
第 8 章/cms4py_first_generation/app/db/async_pydal/row.py
"""

import json
```

```
from datetime import datetime


class Row:
    """
该对象用于封装 aioMySQL 的查询结果
    """

    def __init__(self, field_names, raw_data) -> None:
        super().__init__()
        self._raw_data = raw_data
        self._row_data = {}
        self._field_names = field_names
        for k, v in enumerate(field_names):
            self._row_data[v] = raw_data[k]

    @property
    def field_names(self):
        return self._field_names

    @property
    def raw_data(self):
        return self._raw_data

    @property
    def row_data(self):
        return self._row_data

    def as_dict(self):
        new_dict = {}
        for k, v in self._row_data.items():
            new_dict[k] = v if not isinstance(v, datetime) else str(v)
        return new_dict

    def set(self, key, value):
        self._row_data[key] = value

    def __repr__(self):
        return self.__str__()

    def __str__(self):
        return self._row_data.__str__()

    def __getitem__(self, item):
        return self._row_data[item]

    def __getattr__(self, item):
        return self.__getitem__(item)

    def __contains__(self, item):
```

```
        return self._row_data.__contains__(item)


class Rows:
    """
该对象用于封装 aioMySQL 的查询结果
    """

    def __init__(self, description, raw_data) -> None:
        super().__init__()
        self._rows = []
        self._field_names = []
        self._raw_data = raw_data
        for f in description:
            self._field_names.append(f[0])
        for r in raw_data:
            self._rows.append(Row(self._field_names, r))

    @property
    def rows(self):
        return self._rows

    def as_list(self) -> [dict]:
        """
        Translate to a list contains rows with type of dict
        Returns:
        """
        result = []
        for r in self.rows:
            result.append(r.as_dict())
        return result

    def as_json(self):
        return json.dumps(self.as_list())

    @property
    def field_names(self):
        return self._field_names

    @property
    def raw_data(self):
        return self._raw_data

    def first(self):
        return self.rows[0] if self.rows.__len__() > 0 else None

    def count(self):
        return len(self.rows)

    def __repr__(self):
```

```
        return self.__str__()

    def __str__(self):
        return self._rows.__str__()

    def __getitem__(self, key):
        return self.rows[key]

    def __iter__(self):
        return self.rows.__iter__()

    def __len__(self):
        return self.count()
```

async_pydal 模块初始化文件__init__.py 的源码如下：

```
"""
第 8 章/cms4py_first_generation/app/db/async_pydal/__init__.py
"""

from .dal import AsyncDAL
from .cursor import PyDALCursor
from .connection import PyDALConnection
from .field import Field
from .table import Table
from .row import Row
```

为了方便在 Action 中使用数据库，需要写一个 Db 类用于初始化与管理数据库，代码如下：

```
"""
第 8 章/cms4py_first_generation/app/db/__init__.py
"""

from app.db.async_pydal.dal import AsyncDAL
from pydal import Field


class Db:
    __instance = None

    @staticmethod
    async def get_instance() -> "Db":
        if not Db.__instance:
            Db.__instance = Db()
            await Db.__instance._async_init()
        return Db.__instance

    async def _async_init(self):
```

```
        self._async_pydal = await AsyncDAL.create(
            "127.0.0.1", "root", "rootpw", "mydb"
        )
        self._async_pydal.define_table(
            "student",
            Field("name"),
            Field("age")
        )
        pass

    @property
    def async_pydal(self):
        return self._async_pydal

    pass
```

将数据库配置为如图 8-22 所示。

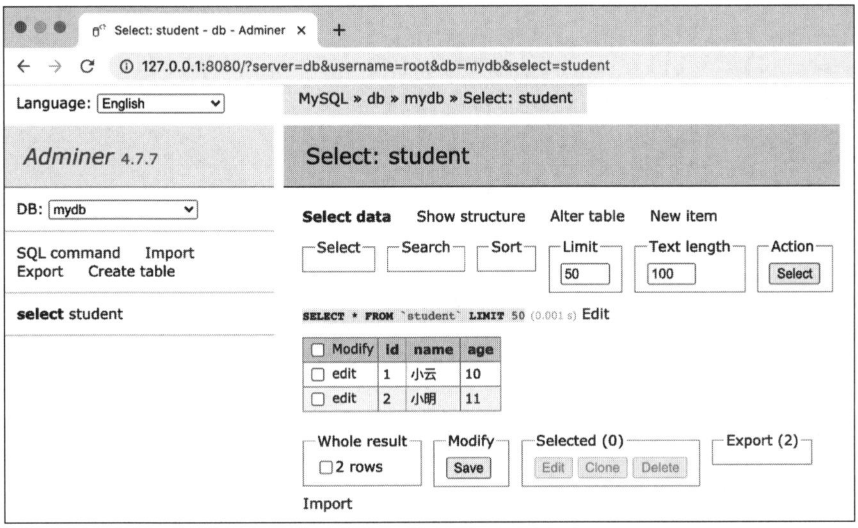

图 8-22　数据库结构与数据示例图

接下来创建一个 Controller 并命名为 students，其代码如下：

```
"""
第 8 章/cms4py_first_generation/app/controllers/students.py
"""

from app.db import Db


async def index(req, res):
    #获取数据管理器实例
    db_manager = await Db.get_instance()
    #通过连接池建立一个连接
```

```
    conn = await db_manager.async_pydal.acquire()
    # 获取数据库对象用于操作数据库
    db = await conn.cursor()
    # 查询数据库
    students = await db(db.student.id > 0).select()
    # 关闭数据库对象
    await db.close()
    # 释放连接
    await db_manager.async_pydal.release(conn)
    await res.render("students.html", students = students)
```

对应的模板文件源码如下：

```
<!-- 第8章/cms4py_first_generation/app/views/students.html -->

<!DOCTYPE html>
<html lang="en">
<head>
<meta charset="UTF-8">
<title>所有学生</title>
</head>
<body>
<table border="1">
<thead>
<tr>
<th>id</th>
<th>名字</th>
<th>年龄</th>
</tr>
</thead>
<tbody>
    {% for s in students %}
<tr>
<td>{{ s.id }}</td>
<td>{{ s.name }}</td>
<td>{{ s.age }}</td>
</tr>
    {% endfor %}
</tbody>
</table>
</body>
</html>
```

页面渲染效果如图8-23所示。

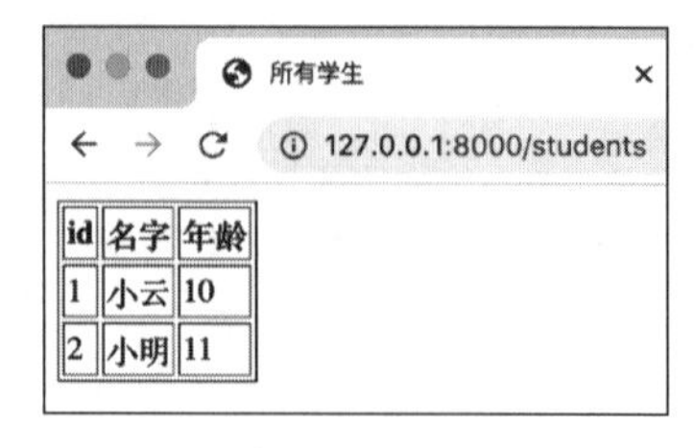

图8-23　数据库渲染后的效果

可以看出，在 Action 中与数据库使用相关的代码用法稍微有点复杂，为了便于后续的开发工作，需要将 Action 进行封装，代码如下：

```
"""
第 8 章/cms4py_first_generation/app/db/action_with_db.py
"""

from cms4py.utils.log import Cms4pyLog
from app.db import Db


class ActionWithDb:

    async def execute(self, req, res):
        raise NotImplementedError()

    async def __call__(self, *args, **kwargs):
        #获取数据管理器实例
        db_manager = await Db.get_instance()
        #通过连接池建立一个连接
        conn = await db_manager.async_pydal.acquire()
        #获取数据库对象,用于操作数据库
        self.db = await conn.cursor()

        #无论在 execute 中发生了什么错误,总要执行释放连接的操作
        err = None
        try:
            await self.execute(*args, **kwargs)
        except BaseException as e:
            err = e
            Cms4pyLog.get_instance().error(e)

        #关闭数据库对象
        await self.db.close()
        #释放连接
        await db_manager.async_pydal.release(conn)

        if err:
            #如果在执行 execute 过程中发生了错误,则将此错误抛给 ASGI
            raise err
```

然后使用与数据库相关的 Action 功能，其写法如下：

```
"""
第 8 章/cms4py_first_generation/app/controllers/students.py
"""

from app.db.action_with_db import ActionWithDb
```

```
class index(ActionWithDb):

    async def execute(self, req, res):
        db = self.db
        students = await db(db.student.id > 0).select()
        await res.render("students.html", students = students)
```

实现同样的功能，在这个 Action 中操作数据库却只用一行代码，不必关心与数据库管理相关的逻辑，因为该逻辑在父类中已经完成了。只要封装得好，代码重用率就高，再复杂的代码也可以用最简单的方式调用。

框架层写了那么多代码就是为了能够在应用层只写这一行代码，所以有人说过，不会偷懒的程序员不是好程序员。为了满足身体的懒惰而让大脑付出了百倍的努力，这种懒惰带来的是生产力的革命。

8.16 集成 Socket.IO

在 cms4py 框架中，设计将所有与 Socket.IO 相关的业务逻辑代码放在 sio 目录下，如图 8-24 所示。

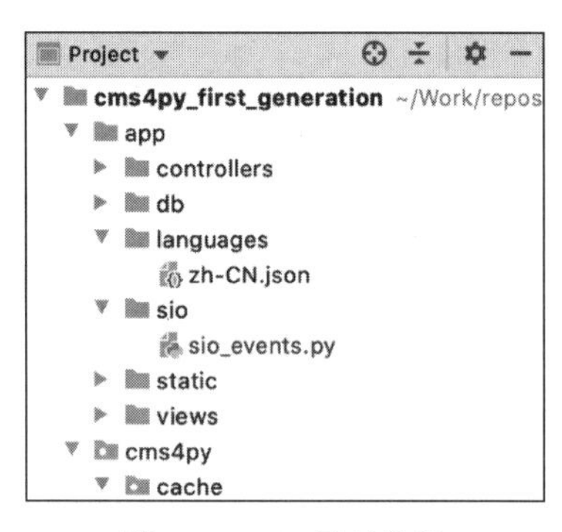

图 8-24 sio 目录位置

为了将此目录的位置设计为可配置，需要在 config.py 文件中添加常量 SOCKET_IO_DIR_NAME，代码如下：

```
"""
第 8 章/cms4py_first_generation/config.py
"""

# 此处省略部分代码

# Socket.IO 文件所在目录
SOCKET_IO_DIR_NAME = "sio"
```

接下来设计 Socket.IO 的支持功能，代码如下：

```
"""
第 8 章/cms4py_first_generation/cms4py/socketio/__init__.py
"""

import socketio, os, config, importlib

sio = socketio.AsyncServer(async_mode = 'asgi')
sio_asgi_app = socketio.ASGIApp(sio)

async def load_sio_files():
    """
加载 sio 文件
    :return:
    """
    sio_root = os.path.join(
        config.APP_DIR_NAME, config.SOCKET_IO_DIR_NAME
    )
    # 从 sio 文件目录中列出所有文件
    sio_files = os.listdir(sio_root)
    for f in sio_files:
        if f.endswith(".py"):
            # 去掉文件后缀名
            f_name = f[:-3]
            # 拼接成模块名
            module_name = "{}.{}.{}".format(
                config.APP_DIR_NAME,
                config.SOCKET_IO_DIR_NAME,
                f_name
            )
            # 导入模块
            importlib.import_module(module_name)
```

函数 load_sio_files 需要在启动时被执行，所以将其调用代码添加到 ASGI 的 lifespan 中，修改 handle_lifespan 函数，代码如下：

```
"""
第 8 章/cms4py_first_generation/cms4py/handlers/lifespan_handler.py
"""
from cms4py.utils.log import Cms4pyLog
from cms4py.socketio import load_sio_files


async def handle_lifespan(scope, receive, send):
    while True:
        # 不断读取数据
        message = await receive()
        # 如果读取消息类型为 lifespan.startup,则进行初始化操作
        if message['type'] == 'lifespan.startup':
            await load_sio_files()
```

```
            # 在初始化完成后，向 ASGI 环境发送启动完成消息
            await send({'type': 'lifespan.startup.complete'})
            Cms4pyLog.get_instance().info("Server started")
        # 如果读取消息类型为 lifespan.shutdown，则进行收尾工作
        elif message['type'] == 'lifespan.shutdown':
            # 在收尾工作结束后，向 ASGI 环境发送收尾完成消息
            await send({'type': 'lifespan.shutdown.complete'})
            Cms4pyLog.get_instance().info("Server stopped")
            break
```

另外还需要在 application 函数中接入 Socket.IO 的支持功能，代码如下：

```
"""
第 8 章/cms4py_first_generation/cms4py/__init__.py
"""

from cms4py.handlers import lifespan_handler
from cms4py.handlers import error_pages
from cms4py.utils.log import Cms4pyLog
from cms4py.handlers import static_file_handler
from cms4py.handlers import dynamic_handler
from cms4py.socketio import sio_asgi_app


async def application(scope, receive, send):
    # 获取请求类型
    request_type = scope['type']

    # 如果是 http 类型的请求，则由该程序段处理
    if request_type == 'http':
        # 如果路径以 /socket.io 开始，则使用 socket.io app 进行处理
        if scope['path'].startswith("/socket.io"):
            return await sio_asgi_app(scope, receive, send)

        data_sent = await dynamic_handler.handle_dynamic_request(
            scope, receive, send
        )

        # 如果经过动态请求处理程序后未发送数据，则尝试使用静态文件请求
        # 处理程序进行处理
        if not data_sent:
            data_sent = await static_file_handler.handle_static_file_request(
                scope, send
            )

        # 如果静态文件处理程序未发送数据，则意味着找不到文件，此时应该
        # 向浏览器发送 404 页面
        if not data_sent:
            # 对于未被处理的请求，均向浏览器发送 404 错误
```

```
                await error_pages.send_404_error(scope, receive, send)
        # 如果是 websocket,则使用 socket.io app 处理
        elif request_type == 'websocket':
            return await sio_asgi_app(scope, receive, send)
        # 如果是生命周期类型的请求,则由该程序段处理
        elif request_type == 'lifespan':
            await lifespan_handler.handle_lifespan(scope, receive, send)
        else:
            Cms4pyLog.get_instance().warning("Unsupported ASGI type")
```

接下来写一段程序对 cms4py 的 Socket.IO 支持功能进行测试,代码如下:

```
"""
第 8 章/cms4py_first_generation/app/sio/sio_events.py
"""

import time, asyncio
from cms4py.socketio import sio


async def echo_task(sid):
    for i in range(1, 6):
        # 向浏览器端发送消息
        await sio.emit(
            "echo",
            f"[{time.strftime('%X')}] Count: {i}",
            sid
        )
        # 等待 1s
        await asyncio.sleep(1)


@sio.event
async def connect(sid, environ):
    """
连接成功
    """
    asyncio.create_task(echo_task(sid))
```

连接该服务器的 html 文件的代码如下:

```
<!-- 第 8 章/cms4py_first_generation/app/static/sio_echo.html -->

<!DOCTYPE html>
<html lang="en">
<head>
<meta charset="UTF-8">
<title>Title</title>
<script src="libs/socket.io.js"></script>
```

```
</head>
<body>
<script>
    (function(){
        let socket = io();
        socket.on("echo",e=>console.log(e))
    })();
</script>
</body>
</html>
```

在启动服务器之前，还需要安装的依赖项有 python-socketio 和 websockets，启动服务器之后通过地址 http://127.0.0.1:8000/sio_echo.html 访问，效果如图 8-25 所示。

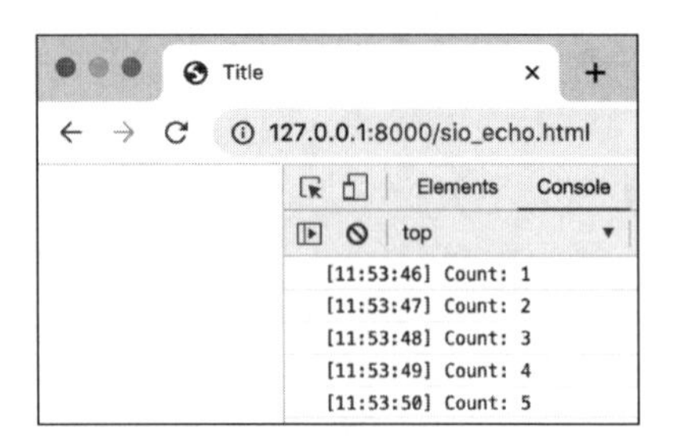

图 8-25 cms4py 集成 Socket.IO 效果测试

8.17 支持 WSGI

cms4py 框架被设计为集大成者，所以应该兼容所有现有技术，包括 WSGI。幸运的是不必动手写 WSGI 的支持代码，Django 团队已经开发了 asgiref 用于支持 WSGI。

下面以一个简单的示例来讲解如何集成 WSGI 应用，代码如下：

```
"""
第 8 章/cms4py_first_generation/cms4py/__init__.py
"""

from cms4py.handlers import lifespan_handler
from cms4py.handlers import error_pages
from cms4py.utils.log import Cms4pyLog
from cms4py.handlers import static_file_handler
from cms4py.handlers import dynamic_handler
from cms4py.socketio import sio_asgi_app
from asgiref.wsgi import WsgiToAsgi


#定义 WSGI 程序
def wsgi(environ, start_response):
    status = '200 OK'
```

```
    output = b'Hello WSGI in WSGI!'

    response_headers = [('Content-type', 'text/plain'),
                        ('Content-Length', str(len(output)))]
    start_response(status, response_headers)

    return [output]


#将 WSGI 包装为 ASGI
wsgi_to_asgi_app = WsgiToAsgi(wsgi)


async def application(scope, receive, send):
    #获取请求类型
    request_type = scope['type']

    #如果是 http 类型的请求,则由该程序段处理
    if request_type == 'http':
        #如果路径以 /socket.io 开始,则使用 socket.io app 进行处理
        if scope['path'].startswith("/socket.io"):
            return await sio_asgi_app(scope, receive, send)

        #如果路径以 /wsgi 开始,则由 asgiref 将请求转交给 wsgi 程序进行处理
        if scope['path'].startswith("/wsgi"):
            return await wsgi_to_asgi_app(scope, receive, send)

        data_sent = await dynamic_handler.handle_dynamic_request(
            scope, receive, send
        )

        #如果经过动态请求处理程序后未发送数据,则尝试使用静态文件请求
        #处理程序进行处理
        if not data_sent:
            data_sent = await static_file_handler.handle_static_file_request(
                scope, send
            )

        #如果静态文件处理程序未发送数据,则意味着找不到文件,此时应该
        #向浏览器发送 404 页面
        if not data_sent:
            #对于未被处理的请求,均向浏览器发送 404 错误
            await error_pages.send_404_error(scope, receive, send)
    #如果是 websocket,则使用 socket.io app 处理
    elif request_type == 'websocket':
        return await sio_asgi_app(scope, receive, send)
    #如果是生命周期类型的请求,则由该程序段处理
    elif request_type == 'lifespan':
        await lifespan_handler.handle_lifespan(scope, receive, send)
```

```
    else:
        Cms4pyLog.get_instance().warning("Unsupported ASGI type")
```

运行效果如图 8-26 所示。

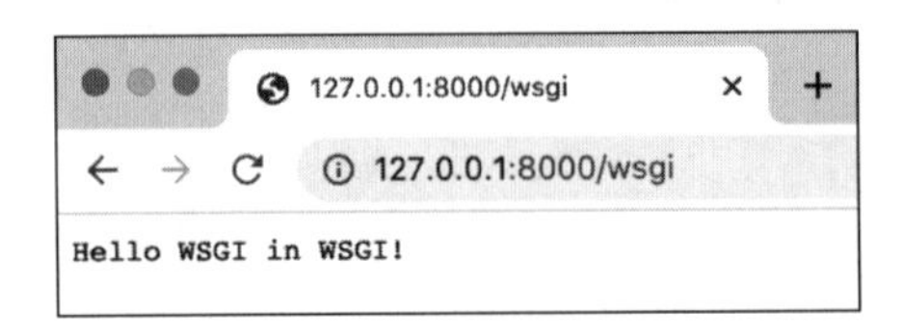

图 8-26 集成 WSGI 效果

8.18 部署在 Apache 服务器后端

虽然 cms4py 所集成的功能已经相当强大，但是在实际生产服务器上，还是倾向于将它部署在 Apache 后端，主要原因有以下几个方面：

(1) Apache 有强大的静态文件处理能力，并且更加稳定、高效，已经经历了几十年的考验。

(2) Apache 集成了所有 Web 服务器应有的功能(ssl、代理与反向代理、虚拟主机、虚拟目录等)，而每个功能的开发都需要巨大的工作量，没必须重新开发。就算能开发，所开发出来的功能也不一定比 Apache 更稳定且高效，实在没有必要浪费时间，所以直接使用 Apache 即可。

(3) Apache 可以方便地集成其他编程语言，如：PHP、Java 等。一个大型网站往往不只是使用某一门编程语言进行开发，根据实际需求，不同的场景将选择最适合的编程语言进行实现，使用 Apache 当网关能够更方便地集成所有其他 Web 开发技术。

接下来我们选择已经集成了 PHP 的 Apache 作为网关，将 cms4py 部署到其后端，并将目录 app/static/static_files 作为静态文件目录交由 Apache 处理访问请求。

Docker 环境配置文件结构如图 8-27 所示。

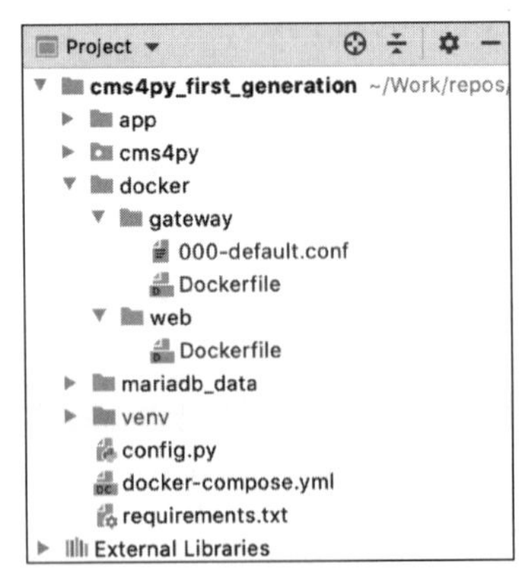

图 8-27 Docker 环境配置文件结构

Docker compose 配置文件内容如下：

```
#第 8 章/cms4py_first_generation/docker-compose.yml

version: '3'

services:

  gateway:
    build: docker/gateway
    ports:
      - 80:80
    volumes:
      #配置 Apache 服务器的根目录
      - "./app/static:/var/www/html"

  web:
    build: docker/web
    #工作目录为 /opt/cms4py
    working_dir: /opt/cms4py
    volumes:
      #将当前目录映射到到容器的 /opt/cms4py 目录
      - ".:/opt/cms4py"
    ports:
      - "8000:8000"
    #将服务器启动在 8000 端口上
    command: uvicorn --host 0.0.0.0 --port 8000 cms4py:application

  db:
    #数据库镜像使用 mariadb
    image: mariadb
    restart: always
    environment:
      #配置 root 用户,密码为 rootpw
      MYSQL_ROOT_PASSWORD: rootpw
    ports:
      #将端口映射到本机
      - 3306:3306
    volumes:
      #将数据库存储地址映射到当前目录下的 mariadb_data 目录
      - "./mariadb_data:/var/lib/MySQL"

  adminer:
    #adminer 是一个数据库管理系统
    image: adminer
    restart: always
    ports:
      #将 adminer 的端口映射到本机,便于访问
      - 8080:8080
```

gateway 服务中的 Docker 环境构建配置文件内容如下：

```
#第 8 章/cms4py_first_generation/docker/gateway/Dockerfile

FROM php:apache

#启用重写模块
RUN a2enmod rewrite
#启用 http 代理模块
RUN a2enmod proxy_http
#启用 FastCGI 代理模块
RUN a2enmod proxy_fcgi
#启用 WebSocket 代理模块
RUN a2enmod proxy_wstunnel


#用自定义的配置文件替换 Apache 配置文件
COPY 000-default.conf /etc/apache2/sites-enabled
```

gateway 服务中的 Apache 配置文件内容如下：

```
#第 8 章/cms4py_first_generation/docker/gateway/000-default.conf

<VirtualHost *:80>
    ServerAdmin webmaster@localhost
    DocumentRoot /var/www/html

    ErrorLog ${APACHE_LOG_DIR}/error.log
    CustomLog ${APACHE_LOG_DIR}/access.log combined

    RewriteEngine On

    #Websocket 代理 W
    RewriteCond %{REQUEST_URI} ^/socket.io              [NC]
    RewriteCond %{QUERY_STRING} transport=websocket    [NC]
    RewriteRule /(.*)        ws://web:8000/$1 [P,L]

    #忽略对 /static_files 路径及其子路径的请求代理
    ProxyPass /static_files !
    ProxyPassReverse /static_files !

    #将其他所有请求代理到 cms4py 服务器
    ProxyPass            / http://web:8000/
    ProxyPassReverse            / http://web:8000/
</VirtualHost>
```

Web 服务中的 Docker 环境构建配置文件内容如下：

```
#第 8 章/cms4py_first_generation/docker/web/Dockerfile

FROM python:3-slim
```

```
#使用清华镜像安装项目依赖项
RUN pip3 install -i https://pypi.tuna.tsinghua.edu.cn/simple \
    asgiref Jinja2 python-socketio uvicorn websockets \
    pydal aioMySQL
```

8.19 技术总结

cms4py 框架集成了大部分常用的 Python Web 开发技术，甚至还对 WSGI 这种旧时代技术提供支持，是一个非常实用的技术框架。

技术框架的设计往往要比应用层业务逻辑的设计花更多的心思。一个网站项目上线后往往会运行十年以上，在功能更新迭代过程中，业务逻辑会随着时代的发展不断地变化，为了适应新的业务逻辑，不得不替换某些技术，甚至对架构进行重构与迁移。

一个好的框架要充分考虑技术迁移的成本（包括时间成本），争取做到可随时根据需求更换技术模块。若能做到这一点，你便可以提升为架构师了。

作为架构师，不仅要考虑架构的强大与灵活，还要考虑设计一种优秀的编码规范用于约束业务逻辑开发者，从而进一步降低框架使用者的技术门槛。在 cms4py 中，设计了完美的 MVC 结构，使用者无须编写路由规则、无须了解数据库连接机制及连接池机制、无须掌握缓存机制、无须理解热更新与优化技术，甚至无须理解什么是 MVC，只须按照 cms4py 的约定去编写控制器、渲染页面、使用现成的 ActionWithDb 类操作数据库，即可编写出高性能的 Web 应用。

希望通过本章的学习与练习，你能够成为一名合格的架构师。

第9章 房屋直租系统项目实例

在第8章中，我们亲手设计并实现了一个全栈框架 cms4py，将众多最新的技术集成在了一起，功能相当强大。为了证明 cms4py 确实强大易用，本章将使用 cms4py 框架开发一个实际项目——房屋直租系统项目。

9.1 制订需求

在 Web 项目开发中，除了框架层解决大多数共性问题之外，还有一些共性问题需要解决，例如：用户系统、权限控制等。虽然每个网站设计或者结构各有不同，但是都会有用户系统，所以笔者在设计这个实战项目时尽最大可能去实践通用的思想。

制订需求功能如下：

（1）用户系统。包括用户注册、登录、权限控制等。

（2）管理系统。管理员可以登录管理系统后台，管理用户及配置网站功能等。

（3）房源分页显示功能。

（4）用户发布房源功能。

（5）用户搜索房源功能。

9.2 技术选型

1. 服务器主机

服务器主机选择云主机。其与传统的独立主机相比，优势在于后期可根据网站实际需要提升主机配置。

2. 操作系统

操作系统选择 Linux。目前主流的选择是 Ubuntu、Debian 和 CentOS，也可以根据开发团队最擅长的系统进行选择，但是前提是必须选择能够安装 Docker 的 Linux 系统。Linux 的优势在于免费、高效、稳定、安全。

3. 容器化技术

Docker 是基于 Linux 内核所提供的容器化技术打造的一个平台，其轻量、高效，可将繁杂的项目维护工作抽象出来，实现脚本化、容器化，从而大大降低项目的维护成本。

4. Web 开发技术

Web 开发技术选择 Python。Python 语言灵活易用，在解决复杂的业务逻辑问题上有着不可替代的优势，从而缩短项目开发周期，节省开发成本。搭配上高效的异步 IO 全栈开发框架 cms4py，在提升开发效率的同时也可提升运行效率。

5. 数据库技术

数据库技术选择 MariaDB。MySQL 曾经是开源数据库的领先者，也曾经是世界上使用最广泛的开源数据库，开发者对于 MySQL 数据库的用法最熟悉不过了。但由于 MySQL 的发展前景并不明朗，而 MariaDB 脱胎于 MySQL，所以最好的选择是 MariaDB，所有用法与 MySQL 一致，不增加额外的学习成本。

9.3 配置运行环境

为了便于后期部署、维护与迁移，选择 Docker 作为运行环境，以第 8 章的 cms4py_first_generation 项目为模板进行配置。

该配置的端口映射如图 9-1 所示。

```
Terminal:  Local ×  +
yunp.top@yunptops-MBP HomeSharing % docker-compose ps
        Name                      Command               State           Ports
-----------------------------------------------------------------------------------------
homesharing_adminer_1   entrypoint.sh docker-php-e ...   Up      0.0.0.0:8080->8080/tcp
homesharing_db_1        docker-entrypoint.sh mysqld      Up      0.0.0.0:3306->3306/tcp
homesharing_gateway_1   docker-php-entrypoint apac ...   Up      0.0.0.0:80->80/tcp
homesharing_web_1       uvicorn --host 0.0.0.0 --p ...   Up      0.0.0.0:8000->8000/tcp
```

图 9-1 开发环境端口映射表

可以看出，所有服务均有端口直接对外公开。在生产服务器上，这是很不安全的，尤其是数据库公开端口。最安全的做法是除 gateway 必须公开的 80 端口（HTTPS 服务器还应该公开 443 端口）外，其他所有端口均不公开。

所以生产服务器上的 docker-compose.yml 文件内容应该进行修改，代码如下：

```
#第 9 章/HomeSharing/docker-compose.yml

version: '3'

services:

  gateway:
    build: docker/gateway
    ports:
      - 80:80
    volumes:
      #配置 Apache 服务器的根目录
      - "./app/static:/var/www/html"

  web:
    build: docker/web
```

```
    #工作目录为 /opt/cms4py
    working_dir: /opt/cms4py
    volumes:
      #将当前目录映射到到容器的 /opt/cms4py 目录
      - ".:/opt/cms4py"
    #将服务器启动在 8000 端口上
    command: uvicorn --host 0.0.0.0 --port 8000 cms4py:application

  db:
    #数据库镜像使用 mariadb
    image: mariadb
    restart: always
    environment:
      #配置 root 用户,密码为 rootpw
      MYSQL_ROOT_PASSWORD: rootpw
    volumes:
      #将数据库存储地址映射到当前目录下的 mariadb_data 目录
      - "./mariadb_data:/var/lib/MySQL"

  adminer:
    #adminer 是一个数据库管理系统
    image: adminer
    restart: always
    ports:
      #将 adminer 的端口映射到本机,便于访问
      - 8080:8080
```

其中 adminer 的端口应该保留,这是为了方便开发者后台操作数据库,但是在生产服务器正式运行的时候,应使用命令 docker-compose stop adminer 停止该服务器以杜绝风险。正式运行的服务器端口映射如图 9-2 所示。

```
Terminal:  Local ×  +
yunp.top@yunptops-MBP HomeSharing % docker-compose ps
        Name                       Command                State          Ports
-------------------------------------------------------------------------------------
homesharing_adminer_1   entrypoint.sh docker-php-e ...   Exit 0
homesharing_db_1        docker-entrypoint.sh mysqld      Up      3306/tcp
homesharing_gateway_1   docker-php-entrypoint apac ...   Up      0.0.0.0:80->80/tcp
homesharing_web_1       uvicorn --host 0.0.0.0 --p ...   Up
```

图 9-2 生产环境端口映射表

在 Docker 环境中,每个服务相当于是一个独立的主机,所有的服务组成了一个局域网,虽然有些端口不对外公开,但是不影响局域网内部服务之间的互相访问。

由于数据库服务名为 db,它在局域网中的域名就是 db,所以 Web 服务访问数据库时所配置的域名应为 db。直接在 PyCharm IDE 中运行的 cms4py 服务器因为运行在本机,不属于 Docker 环境的局域网,所以无法访问数据库服务,这就要求在开发环境中公开数据库的 3306 端口,同时还应修改本机 hosts 文件,添加一行文字“127.0.0.1 db”以便将 db 域名解

析到本机。

在 Windows 平台修改的 hosts 文件为 C:\Windows\System32\drivers\etc\hosts，可使用记事本程序打开，如图 9-3 所示。

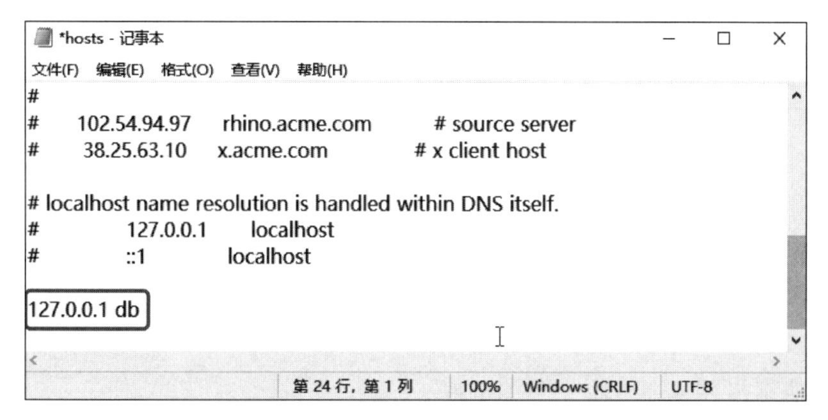

图 9-3　Windows 环境下修改 hosts 文件

Linux 中与 Mac OS X 中的操作方式一样，使用超级用户打开/etc/hosts 文件(命令为 sudo vim /etc/hosts)进行修改，如图 9-4 所示。

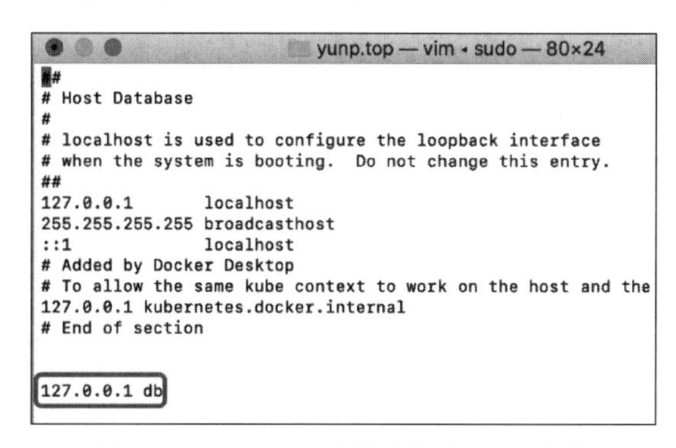

图 9-4　Mac OS X 系统中修改 hosts 文件

9.4　设计数据库结构

使用 Adminer 工具进入数据库管理后台，以 utf8mb4 的编码方式创建名为 home_sharing 的库，如图 9-5 所示。

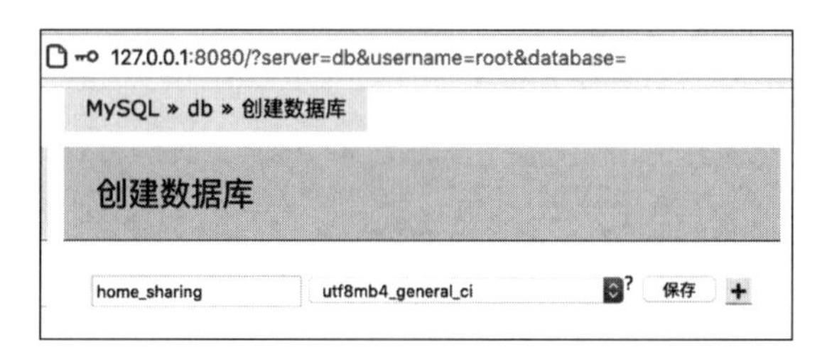

图 9-5　创建数据库

接下来分别设计用户表(auth_user)、用户组表(auth_group)、用户关系表(auth_membership)、上传文件信息表(photo)、房源表(house_res)、房源评论表(house_res_comment)。

用户表(auth_user)结构如表 9-1 所示：

表 9-1　auth_user 表结构

字　段　名	类　　型	说　　明
user_name	字符串	用户名,用于登录网站,唯一字段
user_password	字符串	密码
user_email	字符串	用户邮箱,唯一字段
user_phone	字符串	用户手机号,唯一字段
reg_time	日期	用户注册时间

与用户表结构对应的 SQL 语句如下：

```
-- 第 9 章/HomeSharing/初始数据结构.sql
CREATE TABLE `auth_user` (
  `id` int(11) NOT NULL AUTO_INCREMENT,
  `user_name` tinytext NOT NULL,
  `user_password` varchar(128) NOT NULL,
  `user_email` tinytext NOT NULL,
  `user_phone` varchar(16) NOT NULL,
  `reg_time` datetime NOT NULL DEFAULT '0000 - 00 - 00 00:00:00',
  PRIMARY KEY (`id`),
  UNIQUE KEY `user_phone` (`user_phone`),
  UNIQUE KEY `user_name` (`user_name`) USING HASH,
  UNIQUE KEY `user_email` (`user_email`) USING HASH
) ENGINE = InnoDB DEFAULT CHARSET = utf8mb4;
```

用户组表(auth_group)结构表 9-2 所示。

表 9-2　auth_group 表结构

字　段　名	类　　型	说　　明
role	字符串	组名
description	字符串	组说明

用户组用于控制用户权限。在程序中执行某操作时可检查当前登录的用户是否属于该组,以此控制当前用户是否有权限执行该操作。

与用户组表结构对应的 SQL 语句如下：

```
-- 第 9 章/HomeSharing/初始数据结构.sql
CREATE TABLE `auth_group` (
  `id` int(11) NOT NULL AUTO_INCREMENT,
  `role` tinytext NOT NULL,
  `description` tinytext NOT NULL,
```

```
  PRIMARY KEY (`id`),
  UNIQUE KEY `role` (`role`) USING HASH
) ENGINE = InnoDB DEFAULT CHARSET = utf8mb4;
```

用户与用户组的对应关系是多对多的关系，所以应当另创建一个关系表用于记录用户与用户组的对应关系，关系表(auth_membership)的结构如表 9-3 所示。

表 9-3　auth_membership 表结构

字　段　名	类　　型	说　　明
group_id	引用 auth_group	关联组 id
user_id	引用 auth_user	关联用户 id

与之对应的 SQL 语句如下：

```
-- 第 9 章/HomeSharing/初始数据结构.sql
CREATE TABLE `auth_membership` (
  `id` int(11) NOT NULL AUTO_INCREMENT,
  `group_id` int(11) NOT NULL,
  `user_id` int(11) DEFAULT NULL,
  PRIMARY KEY (`id`),
  KEY `group_id` (`group_id`),
  KEY `user_id` (`user_id`),
  CONSTRAINT `auth_membership_ibfk_1` FOREIGN KEY (`group_id`) REFERENCES `auth_group` (`id`),
  CONSTRAINT `auth_membership_ibfk_2` FOREIGN KEY (`user_id`) REFERENCES `auth_user` (`id`)
) ENGINE = InnoDB DEFAULT CHARSET = utf8mb4;
```

发布房源时需要上传图片，所以应当设计图片的保存机制。因为图片比较大，图片的读写比较耗资源，所以在实际项目中，不建议将图片数据直接保存在数据库中，而是将图片保存在硬盘上，数据库中只保留该图片的路径，从而减轻数据库的压力。

设计存储图片路径的表(photo)结构如表 9-4 所示。

表 9-4　photo 表结构

字　段　名	类　　型	说　　明
photo_name	字符串	图片名称
photo_uri	字符串	访问该图片所用的路径
photo_path	字符串	图片在服务器主机硬盘上的路径
creator	引用 auth_user	上传者 id
creation_time	日期	创建图片的时间

与之对应的 SQL 语句如下：

```
-- 第 9 章/HomeSharing/初始数据结构.sql

CREATE TABLE `photo` (
  `id` int(11) NOT NULL AUTO_INCREMENT,
  `photo_name` tinytext NOT NULL,
  `photo_uri` tinytext NOT NULL,
```

```
  `photo_path` tinytext NOT NULL,
  `creator` int(11) NOT NULL,
  `creation_time` datetime NOT NULL DEFAULT '0000 - 00 - 00 00:00:00',
  PRIMARY KEY (`id`),
  KEY `creator` (`creator`),
  CONSTRAINT `photo_ibfk_1` FOREIGN KEY (`creator`) REFERENCES `auth_user` (`id`)
) ENGINE = InnoDB DEFAULT CHARSET = utf8mb4;
```

房源表(house_res)结构如表 9-5 所示。

表 9-5 house_res 表结构

字段名	类型	说明
res_title	字符串	房源标题
res_content	文本	房源描述
pub_time	日期	房源发布时间
owner_id	引用 auth_user	发布者 id

对应的 SQL 语句如下：

```
-- 第 9 章/HomeSharing/初始数据结构.sql
CREATE TABLE `house_res` (
  `id` int(11) NOT NULL AUTO_INCREMENT,
  `res_title` tinytext NOT NULL,
  `res_content` longtext NOT NULL,
  `pub_time` datetime NOT NULL,
  `owner_id` int(11) NOT NULL,
  PRIMARY KEY (`id`),
  KEY `owner_id` (`owner_id`),
  CONSTRAINT `house_res_ibfk_1` FOREIGN KEY (`owner_id`) REFERENCES `auth_user` (`id`)
) ENGINE = InnoDB DEFAULT CHARSET = utf8mb4;
```

房源应允许评论，所以再设计一个房源评论表(house_res_comment)，结构如表 9-6 所示。

表 9-6 house_res_comment 表结构

字段名	类型	说明
user_id	引用 auth_user	评论者用户 id
house_res_id	引用 house_res	房源 id
comment_content	文本	评论内容
comment_time	日期	发布评论的时间

对应的 SQL 语句如下：

```
-- 第 9 章/HomeSharing/初始数据结构.sql

CREATE TABLE `house_res_comment` (
```

```
  `id` int(11) NOT NULL AUTO_INCREMENT,
  `user_id` int(11) NOT NULL,
  `house_res_id` int(11) NOT NULL,
  `comment_content` text NOT NULL,
  `comment_time` datetime NOT NULL,
  PRIMARY KEY (`id`),
  KEY `user_id` (`user_id`),
  KEY `house_res_id` (`house_res_id`),
  CONSTRAINT `house_res_comment_ibfk_1` FOREIGN KEY (`user_id`) REFERENCES `auth_user` (`id`),
  CONSTRAINT `house_res_comment_ibfk_2` FOREIGN KEY (`house_res_id`) REFERENCES `house_res` (`id`)
) ENGINE = InnoDB DEFAULT CHARSET = utf8mb4;
```

实际上这个数据库结构设计是一个通用的CMS(Content Management System,内容管理系统)结构,该结构可以轻易修改为博客、论坛、新闻站、企业门户等。其实cms4py之所以命名为cms4py,初衷也是这个,就是打造一个通用的快速建站工具系统,笔者再次呼吁有志青年以练手之名加入这个开源项目中,一起打造一个影响世界的内容管理系统。

9.5 实现用户系统

首先需要编写与数据库相关的定义,为了便于后期代码维护,将为每个数据表编写单独的配置文件。

表auth_user对应的配置文件代码如下:

```
"""
第 9 章/HomeSharing/app/db/table_auth_user.py
"""

from pydal import Field
from pydal.validators import CRYPT


def define_table(db):
    db.define_table(
        "auth_user",
        Field("user_name"),
        Field("user_password", requires = CRYPT()),
        Field("user_email"),
        Field("user_phone"),
        Field("reg_time", type = "datetime"),
    )
    pass
```

表auth_group对应的配置文件代码如下:

```
"""
第 9 章/HomeSharing/app/db/table_auth_group.py
```

```
"""

from pydal import Field


def define_table(db):
    db.define_table(
        "auth_group",
        Field('role'),
        Field('description'),
    )
    pass
```

表 auth_membership 对应的配置文件代码如下：

```
"""
第 9 章/HomeSharing/app/db/table_auth_membership.py
"""

from pydal import Field


def define_table(db):
    db.define_table(
        "auth_membership",
        Field('group_id', "reference user_group"),
        Field('user_id', "reference user")
    )
```

在初始化 db 模块时必须执行这 3 个函数，以对数据库进行初始化定义，代码如下：

```
"""
第 9 章/cms4py_first_generation/app/db/__init__.py
"""

from app.db.async_pydal.dal import AsyncDAL
from . import table_auth_group
from . import table_auth_membership
from . import table_auth_user


class Db:
    __instance = None

    @staticmethod
    async def get_instance() -> "Db":
        if not Db.__instance:
            Db.__instance = Db()
            await Db.__instance._async_init()
```

```
        return Db.__instance

    async def _async_init(self):
        self._async_pydal = await AsyncDAL.create(
            "db", "root", "rootpw", "home_sharing"
        )

        table_auth_user.define_table(self._async_pydal)
        table_auth_group.define_table(self._async_pydal)
        table_auth_membership.define_table(self._async_pydal)
        pass

    @property
    def async_pydal(self):
        return self._async_pydal

    pass
```

接下来修改 ActionWithDb 类，以加入自动保存数据的功能，代码如下：

```
"""
第 9 章/cms4py_first_generation/app/db/action_with_db.py
"""

from cms4py.utils.log import Cms4pyLog
from app.db import Db


class ActionWithDb:

    async def execute(self, req, res):
        raise NotImplementedError()

    async def __call__(self, *args, **kwargs):
        #获取数据管理器实例
        db_manager = await Db.get_instance()
        #通过连接池建立一个连接
        conn = await db_manager.async_pydal.acquire()
        #自动提交数据以便更改
        await conn.autocommit(True)
        #获取数据库对象,用于操作数据库
        self.db = await conn.cursor()

        #无论在 execute 中发生了什么错误,总要执行释放连接的操作
        err = None
        try:
            await self.execute(*args, **kwargs)
        except BaseException as e:
            err = e
```

```
            Cms4pyLog.get_instance().error(e)

        #关闭数据库对象
        await self.db.close()
        #释放连接
        await db_manager.async_pydal.release(conn)
        if err:
            #如果在执行 execute 过程中发生了错误,将此错误抛给 ASGI
            raise err
```

与数据库相关的准备工作已经完成,下面准备基础模板文件。在该项目中,为了快速搭建界面,将采用前端技术 Bootstrap。前端依赖项文件结构如图 9-6 所示。

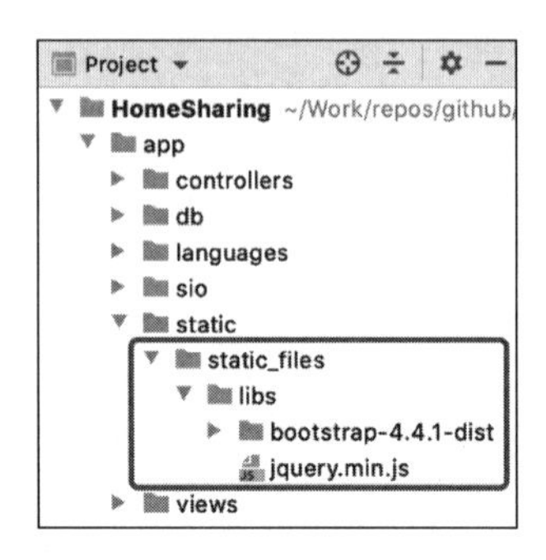

图 9-6　前端依赖项文件结构

网站页面程序有一个特点,同一个网站的所有页面主框架是一样的,这意味着每个页面都有一部分重复的代码。在 Jinja2 中,模板是可以继承的,只需要写一个框架模板代码作为所有页面的根模板即可,此根模板文件代码如下:

```
<!-- 第 9 章/HomeSharing/app/views/layout.html -->
<!DOCTYPE html>
<html lang="en">
<head>
<meta charset="UTF-8">
<title>{{title}}</title>
<link rel="stylesheet" href="/static_files/libs/bootstrap-4.4.1-dist/css/bootstrap.min.css">

<script src="/static_files/libs/jquery.min.js"></script>
<script src="/static_files/libs/bootstrap-4.4.1-dist/js/bootstrap.bundle.min.js"></script>
</head>
<body>
<nav class="navbar navbar-expand-sm navbar-light bg-light">
<a class="navbar-brand" href="#">房屋直租</a>
<button class="navbar-toggler" type="button" data-toggle="collapse" data-target="#navbarSupportedContent"
        aria-controls="navbarSupportedContent" aria-expanded="false" aria-label="Toggle navigation">
```

```
< span class = "navbar - toggler - icon"></span >
</button >

< div class = "collapse navbar - collapse" id = "navbarSupportedContent">
< ul class = "navbar - nav mr - auto">
< li class = "nav - item">
< a class = "nav - link" href = "#">房源</a >
</li >
</ul >
< form class = "form - inline my - 2 my - lg - 0">
< input class = "form - control mr - sm - 2" type = "search" placeholder = "关键字" aria - label
 = "Search">
< button class = "btn btn - outline - success my - 2 my - sm - 0" type = "submit">搜索</button >
</form >
< ul class = "navbar - nav">
            {% if 'current_user' in session and session['current_user'] %}
            {% set user = session['current_user'] %}
< li class = "nav - item dropdown">
< a class = "nav - link dropdown - toggle" href = "#" id = "navbarDropdown" role = "button" data
 - toggle = "dropdown"
                aria - haspopup = "true" aria - expanded = "false">
                 {{ user['user_name'] }}
</a >
< div class = "dropdown - menu dropdown - menu - right" aria - labelledby = "navbarDropdown">
< a class = "dropdown - item" href = "/user/profile">我的信息</a >
< div class = "dropdown - divider"></div >
< a class = "dropdown - item" href = "/user/logout">退出</a >
</div >
</li >
            {% else %}
< li class = "nav - item">
< a class = "nav - link" href = "/user/login">登录</a >
</li >
< li class = "nav - item">
< a class = "nav - link" href = "/user/register">注册</a >
</li >
            {% endif %}
</ul >
</div >
</nav >
< div class = "container">
    {% block body %}{% endblock %}
</div >
</body >
</html >
```

接下来实现注册页面模板文件 register. html，代码如下：

```
<!-- 第 9 章/HomeSharing/app/views/user/register.html -->

{ % extends "layout.html" % }
{ % block body % }
< div class = "container" id = "vueapp">
< div class = "card" style = "margin:20px auto 20px auto;width: 500px;">
< div class = "card - header font - weight - bold">
注册
</div >
< div class = "card - body">
< form id = "form - register" style = "display: block;" enctype = "multipart/form - data" method
 = "post">
< div style = "text - align:center" class = "alert - msg text - danger">
                    { % if alert_msg % } {{ alert_msg }} { % endif % }
</div >
< table class = "table table - borderless">
< tbody >
< tr >
< td class = "font - weight - bold">用户名</td >
< td >
< input type = "text" name = "login_name" required class = "form - control">
</td >
</tr >
< tr >
< td class = "font - weight - bold">邮箱</td >
< td >
< input type = "email" name = "email" required class = "form - control">
</td >
</tr >
< tr >
< td class = "font - weight - bold">手机号</td >
< td >
< input type = "text" name = "phone" required class = "form - control">
</td >
</tr >
< tr >
< td class = "font - weight - bold">密码</td >
< td >
< input type = "password" name = "password" required class = "form - control">
</td >
</tr >
< tr >
< td class = "font - weight - bold">密码确认</td >
< td >
< input type = "password" name = "password_confirm" required class = "form - control">
</td >
</tr >
</tbody >
</table >
```

```
<div>
<input type="submit" class="btn btn-outline-success" value="注册">
<a class="text-success" href="/user/login" style="float:right;">
已有账号?点此登录
</a>
</div>
</form>
</div>
</div>
</div>
<script src="/static_files/user_register.js"></script>
{% endblock %}
```

与该模板文件匹配的还有一个 user_register.js 文件，其用途是在提交表单之前判断密码确认是否一致，其代码如下：

```
//第 9 章/HomeSharing/app/static/static_files/user_register.js

(function () {

    var alertMsgDiv = document.querySelector(".alert-msg");

    document.querySelector("#form-register").onsubmit = function (e) {
        if (this['password_confirm'].value != this['password'].value) {
            e.preventDefault(); //确认密码不一致时阻止提交
            alertMsgDiv.innerHTML = "确认密码不一致";
        }
    };
})();
```

与用户注册功能相对应的 Python 业务逻辑代码如下：

```
"""
第 9 章/HomeSharing/app/controllers/user.py
"""

from app.db.action_with_db import ActionWithDb
from cms4py.http import Request, Response
import datetime, re
from pyMySQL.err import IntegrityError

class register(ActionWithDb):
    async def execute(self, req: Request, res):
        db = self.db
        alert_msg = ""
        if req.method == 'GET':
            await res.render(
                "user/register.html", title="注册",
                alert_msg=alert_msg
```

```
            )
        elif req.method == "POST":
            login_name = req.get_var_as_str(b"login_name")
            email = req.get_var_as_str(b"email")
            phone = req.get_var_as_str(b"phone")
            password = req.get_var_as_str(b"password")
            #将密码进行加密存储
            password = self.db.auth_user.user_password.requires(
                password
            )[0]
            try:
                if (await self.db.auth_user.insert(
                        user_name = login_name, user_email = email,
                        user_phone = phone, user_password = password,
                        reg_time = datetime.datetime.utcnow()
                )):
                    user_record = (
                        await db(
                            db.auth_user.user_name == login_name
                        ).select()
                    ).first()
                    await req.set_session(
                        "current_user", user_record.as_dict()
                    )
                    await res.redirect("/user/profile")
            except IntegrityError as err:
                alert_msg = "注册失败"

                #错误提示信息,如
                #Duplicate entry 'a@a.a' for key 'user_email'
                #该正则的目的是将字段名从错误提示信息中提取出来
                result = re.search(
                    "Duplicate entry .* for key '(\\w+)'",
                    err.args[1]
                )
                error_msg = result[1]
                if error_msg == 'user_email':
                    alert_msg = "邮箱已被占用"
                elif error_msg == 'user_name':
                    alert_msg = "用户名已被占用"
                elif error_msg == 'user_phone':
                    alert_msg = "手机号已被占用"
                await res.render(
                    "user/register.html", title = "注册",
                    alert_msg = alert_msg
                )
            else:
                alert_msg = "注册失败"
                await res.render(
                    "user/register.html", title = "注册",
```

```
                    alert_msg = alert_msg
                )
        else:
            await res.end(b"Method not supported")
```

页面渲染效果如图 9-7 所示。

图 9-7　注册页面渲染效果

注册页面与登录页面可以互相切换，方便用户进行登录及注册操作，登录页面模板源码如下：

```
<!-- 第 9 章/HomeSharing/app/views/user/login.html -->
{ % extends "layout.html" % }
{ % block body % }
< div class = "container">
< div class = "card" style = "width: 500px;margin: 20px auto 20px auto;">
< div class = "card - header">登录</div >
< div class = "card - body">
< form method = "post" enctype = "multipart/form - data">
< div style = "text - align:center" class = "alert - msg text - danger">
                { % if alert_msg % } {{ alert_msg }} { % endif % }
</div >
< table class = "table table - borderless">
< tbody >
< tr >
```

```
<td class = "font - weight - bold">用户名</td>
<td>
<input type = "text" class = "form - control" required name = "login"
                                   placeholder = "用户名/邮箱/手机号">
</td>
</tr>
<tr>
<td class = "font - weight - bold">密码</td>
<td><input type = "password" class = "form - control" required name = "password"></td>
</tr>
</tbody>
</table>
<div>
<div style = "float: right;">
<a href = "/user/register" class = "text - success" style = "display: block;">
没有帐户?点此注册
</a>
</div>
<input type = "submit" class = "btn btn - outline - success" value = "登录">
</div>
</form>
</div>
</div>
</div>
{ % endblock % }
```

登录业务逻辑代码如下：

```
"""
第 9 章/HomeSharing/app/controllers/user.py
"""

from app.db.action_with_db import ActionWithDb
from cms4py.http import Request, Response
import datetime, re
from pyMySQL.err import IntegrityError


class login(ActionWithDb):
    async def execute(self, req: Request, res):
        if req.method == "GET":
            await res.render("user/login.html", title = "登录")
        elif req.method == 'POST':
            login_param = req.get_var_as_str(b"login")
            password = req.get_var_as_str(b"password")

            db = self.db
            user_record = (await db(
                (db.auth_user.user_name == login_param) |
```

```
                (db.auth_user.user_email == login_param) |
                (db.auth_user.user_phone == login_param)
            ).select()).first()
            _next = None
            if user_record:
                transcode = db.auth_user.user_password.requires(password)[0]
                if transcode == user_record.user_password:
                    # 将登录用户写入 session
                    await req.set_session(
                        "current_user", user_record.as_dict()
                    )
                    await res.redirect("/user/profile")
                else:
                    alert_msg = "用户名或密码错误"
                    await res.render(
                        "user/login.html",
                        title = "登录", alert_msg = alert_msg
                    )
        else:
            await res.end(b"Unsupported method")
```

登录页面渲染效果如图 9-8 所示。

图 9-8　登录页面渲染效果

在登录完成后跳转到用户信息(Profile)页面，其模板代码如下：

```
<!-- 第 9 章/HomeSharing/app/views/user/profile.html -->
{ % extends "layout.html" % }
{ % block body % }
    { % if "current_user" in session % }
        { % set user = session['current_user'] % }
<div class="card" style="margin-top:10px">
```

```
<div class="card-header" style="display: flex;flex-direction: row;">
<div>
用户信息
</div>
</div>
<div class="card-body">
<table class="table table-borderless">
<tbody>
<tr>
<td class="font-weight-bold" style="width: 200px;">用户名</td>
<td>{{ user["user_name"] }}</td>
</tr>
<tr>
<td class="font-weight-bold">邮箱</td>
<td>{{ user["user_email"] }}</td>
</tr>
<tr>
<td class="font-weight-bold">手机号</td>
<td>{{ user["user_phone"] }}</td>
</tr>
</tbody>
</table>
</div>
</div>
    {% endif %}
{% endblock %}
```

用户信息页面的业务逻辑代码如下：

```
async def profile(req, res):
    await res.render("user/profile.html", title="用户信息")
```

用户信息页面渲染效果如图 9-9 所示。

图 9-9　用户信息页面渲染效果

在根模板中有定义右上角的用户信息操作模块，在当前用户处于非登录状态下，右上角呈现"注册"与"登录"两个链接。在处于登录状态时将呈现一个下拉菜单用于跳转到用户信息页面及登出，如图 9-10 所示。

图 9-10　用户详情菜单

该菜单中与登录相关的功能业务逻辑代码如下：

```
async def logout(req: Request, res: Response):
    await req.set_session("current_user", None)
    await res.redirect("/user/login")
```

9.6 实现权限系统

在服务器端项目的开发中，权限控制是比较复杂的一个功能，设计不好将对后期项目开发造成巨大不利影响。为了把权限系统设计得易懂且易用，这里将采用装饰器进行控制。

创建一个单独的文件并命名为 auth.py，在其中编写与权限控制相关的装饰器函数，代码如下：

```
"""
第 9 章/HomeSharing/app/db/auth.py
"""

from cms4py.http import Request


async def has_membership(db, request, response, role, user_id = None):
    """
根据用户 id 或者当前登录的用户判断是否属于指定的组
    :param db: PyDALCursor 对象
    :param request:
    :param response:
```

```
    :param role:
    :param user_id:
    :return:
    """
    if not user_id:
        current_user = await request.get_session("current_user")
        user_id = current_user['id'] \
            if (current_user and 'id' in current_user) \
            else None
    if not user_id:
        return False
    return not (await db(
        (db.auth_group.role == role) &
        (db.auth_membership.user_id == user_id) &
        (db.auth_group.id == db.auth_membership.group_id)
    ).isempty())

def require_login():
    """
访问目标 Action,要求用户登录
    :return:
    """

    def outer(func):
        async def wrapper( * args):
            argc = len(args)
            if argc == 3:
                req: Request = args[1]
                res = args[2]
            elif argc == 2:
                req: Request = args[0]
                res = args[1]
            else:
                raise TypeError()
            #如果已登录,则执行目标函数
            if await req.get_session('current_user'):
                return await func( * args)
            #如果未登录,则跳转到登录页面
            else:
                await res.redirect("/user/login")

        return wrapper

    return outer


def require_membership( * roles):
    """
目标 Action 要求用户在指定的组里,目标 Action 必须
是 ActionWithDb 类型
```

```
    :param roles:
    :return:
    """

    def outer(func):
        async def inner(action_with_db, request: Request, response):
            if await request.get_session("current_user"):
                for r in roles:
                    if await has_membership(
                            action_with_db.db, request, response, r
                    ):
                        await func(action_with_db, request, response)
                        return
                await response.end(b"Access Denied")
            else:
                await response.redirect("/user/login")

        return inner

    return outer
```

接下来以要求用户登录为例演示如何使用与权限相关的功能，在与/user/profile 相关的业务逻辑中要求用户登录的代码如下：

```
@auth.require_login()
async def profile(req, res):
    await res.render("user/profile.html", title="用户信息")
```

在为某一 Action 函数加上@auth.require_login()装饰器后，访问该 Action，如果用户处于登录状态，则正常显示页面，否则将跳转至登录页面。

9.7 管理面板

任何系统都应提供与管理相关的功能，否则该系统将不会有价值。登录管理面板可以管理系统中的所有功能。

在实现管理面板之前，需要先在数据库中进行一些初始化的配置工作，以将其中一个用户配置为管理员。

首先注册一个用户名为 admin 的用户，在数据库的截图如图 9-11 所示。

注意观察，这里的 admin 用户的 id 是 13，在你注册完成后也许不是此值。

接下来添加 admin 组，admin 组中的用户是管理员用户，可以管理网站所有功能，添加后在 auth_group 表如图 9-12 所示。

接下来配置 admin 用户与 admin 组的关联，操作如图 9-13 所示。

在此操作中，id 为自增字段，留空即可，group_id 为 admin 组的 id，user_id 为 admin 用户的 id，添加关联完成后，如图 9-14 所示。

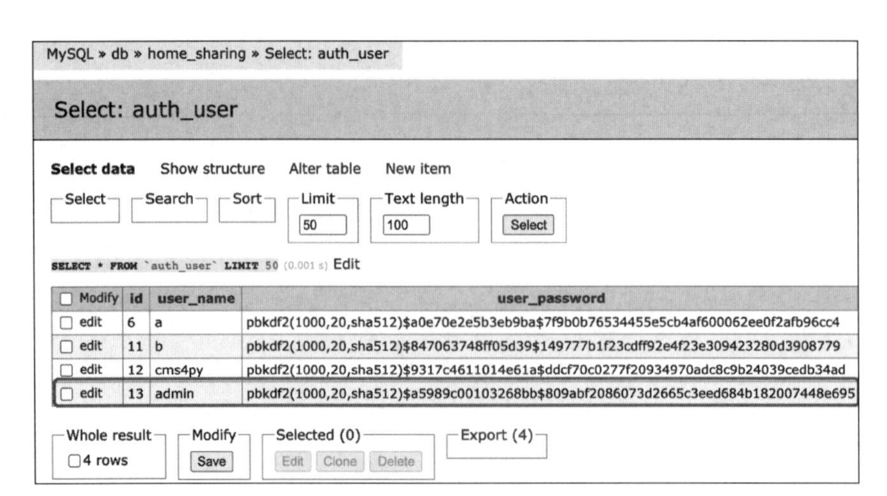

图 9-11　admin 在库中

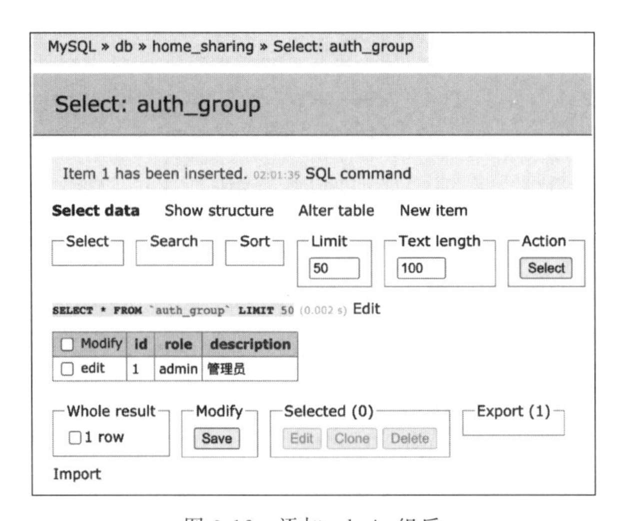

图 9-12　添加 admin 组后

MySQL » db » home_sharing » auth_membership » Insert

Insert: auth_membership

id	Auto Increment	
group_id		1
user_id		13

Save　Save and insert next

图 9-13　添加关联

MySQL » db » home_sharing » Select: auth_membership

Select: auth_membership

Select data　Show structure　Alter table　New item

Select　Search　Sort　Limit 50　Action Select

SELECT * FROM `auth_membership` LIMIT 50 (0.001 s) Edit

Modify	id	group_id	user_id
edit	1	1	13

Whole result 1 row　Modify Save　Selected (0) Edit Clone Delete　Export (1)

Import

图 9-14　admin 用户关联

这一步完成后，admin 用户将拥有 admin 组的权限。

接下来实现管理面板根模板，代码如下：

```
<!-- 第 9 章/HomeSharing/app/views/admin_layout.html -->
<!DOCTYPE html>
<html lang = "en">
<head>
<meta charset = "UTF - 8">
<title>{{title}}</title>
<link rel = "stylesheet" href = "/static_files/libs/bootstrap - 4.4.1 - dist/css/bootstrap.
min.css">

<script src = "/static_files/libs/jquery.min.js"></script>
<script src = "/static_files/libs/bootstrap - 4.4.1 - dist/js/bootstrap.bundle.min.js">
</script>
</head>
<body>
<nav class = "navbar navbar - expand - sm navbar - light bg - light">
<a class = "navbar - brand" href = "/admin"> cms4py 管理系统</a>
<button class = "navbar - toggler" type = "button" data - toggle = "collapse" data - target = "#
navbarSupportedContent"
            aria - controls = "navbarSupportedContent" aria - expanded = "false" aria - label
= "Toggle navigation">
<span class = "navbar - toggler - icon"></span>
</button>

<div class = "collapse navbar - collapse" id = "navbarSupportedContent">
<ul class = "navbar - nav mr - auto">
<li class = "nav - item">
<a class = "nav - link" href = "/admin/users">用户</a>
</li>
<li class = "nav - item">
<a class = "nav - link" href = "/admin/groups">组</a>
</li>
<li class = "nav - item">
<a class = "nav - link" href = "/admin/memberships">关系表</a>
</li>
</ul>
<ul class = "navbar - nav">
            {% if 'current_user' in session and session['current_user'] %}
            {% set user = session['current_user'] %}
<li class = "nav - item dropdown">
<a class = "nav - link dropdown - toggle" href = "#" id = "navbarDropdown" role = "button"
data - toggle = "dropdown"
                    aria - haspopup = "true" aria - expanded = "false">
                     {{ user['user_name'] }}
</a>
<div class = "dropdown - menu dropdown - menu - right" aria - labelledby = "navbarDropdown">
<a class = "dropdown - item" href = "/user/profile">我的信息</a>
```

```
<div class="dropdown-divider"></div>
<a class="dropdown-item" href="/user/logout">登出</a>
</div>
</li>
            {% else %}
<li class="nav-item">
<a class="nav-link" href="/user/login">登录</a>
</li>
<li class="nav-item">
<a class="nav-link" href="/user/register">注册</a>
</li>
            {% endif %}
</ul>
</div>
</nav>
<div class="container">
    {% block body %}{% endblock %}
</div>
</body>
</html>
```

管理面板首页代码如下：

```
<!-- 第 9 章/HomeSharing/app/views/admin/index.html -->
{% extends "admin_layout.html" %}
{% block body %}
欢迎使用 cms4py 管理面板
{% endblock %}
```

对应的业务逻辑代码如下：

```
"""
第 9 章/HomeSharing/app/controllers/admin.py
"""
from ..db.action_with_db import ActionWithDb
from ..db import auth


class index(ActionWithDb):

    @auth.require_membership("admin")
    async def execute(self, req, res):
        await res.render("admin/index.html", title="cms4py 管理面板")
```

如果当前用户没有登录，则跳转到登录页面，如果已登录但无权限，则提示用户没有权限，只有在当前登录用户有权限的情况下才能正常显示该页面。

9.8 呈现关系表

在实际项目中并不需要直接将用户关系表呈现出来，而是将该功能集成到用户权限管理中，本节目的在于讲解表格的渲染及表格分页。

查询关系表的 Action 代码如下：

```
"""
第 9 章/HomeSharing/app/controllers/admin.py
"""
from ..db.action_with_db import ActionWithDb
from ..db import auth


class memberships(ActionWithDb):

    @auth.require_membership("admin")
    async def execute(self, req, res):
        db = self.db
        rows = await db(db.auth_membership.id > 0).select(
            #按 id 反序排列
            orderby = ~db.auth_membership.id
        )
        await res.render(
            "admin/memberships.html", title = "关系表",
            rows = rows
        )
```

对应的模板文件代码如下：

```
<!-- 第 9 章/HomeSharing/app/views/admin/index.html -->
{ % extends "admin_layout.html" % }
{ % block body % }
<div style = "margin - top:1rem;">
<table class = "table table - bordered">
<thead>
<tr>
<th> id </th>
<th> group_id </th>
<th> user_id </th>
</tr>
</thead>
<tbody>
        { % for r in rows % }
<tr>
<td>{{ r.id }}</td>
<td>{{ r.group_id }}</td>
<td>{{ r.user_id }}</td>
```

```
</tr>
            {% endfor %}
</tbody>
</table>
</div>
{% endblock %}
```

渲染效果如图 9-15 所示。

图 9-15　关系表渲染结果

如果数据量较少，可以用这种方式渲染，但如果数据量比较庞大，则应考虑分页呈现。分页呈现数据的代码逻辑比较复杂，而且可适用于很多场景，所以有必要写一个公共组件，以便对数据进行自动分页。

公共组件 data_grid 代码如下：

```
"""
第 9 章/HomeSharing/app/db/data_grid.py
"""

from typing import Awaitable
from cms4py.http import Request, Response
import config
from URLlib.parse import unquote
import math
import inspect
from app.utils import URL


def default_row_render(db, request, response, row, fields):
    html_str = "<tr>"
    for f in fields:
        html_str += f"<td>{row[f]}</td>"
    html_str += "</tr>"
    return html_str


def default_header_render(db, request, response, fields):
    html_str = "<tr>"
```

```
        for f in fields:
            html_str += f"<th>{response.translate(f)}</th>"
        html_str += "</tr>"
        return html_str


def default_foot_render(
        db,
        request: Request,
        response,
        current_page_index,
        paginate, all_count
):
    """
默认底部选择器连接渲染器,算法为根据当前页码向左最多呈现 5 个页码
链接,向右最多呈现 5 个页码链接
    :param db:
    :param request:
    :param response:
    :param current_page_index:
    :param paginate:
    :param all_count:
    :return:
    """
    vars = {}
    for k in request.query_vars:
        vars[k.decode(config.GLOBAL_CHARSET)] = unquote(
            request.get_query_var(k).decode(config.GLOBAL_CHARSET)
        )

    def create_link(page_index, label = None, active = False):
        vars['page_index'] = page_index
        return f"<li class = \"page - item {'active' if active else ''}\">" \
                f" <a " \
                f" class = \"btn - page - number page - link\" " \
                f" href = '{request.path}?{URL.dict_to_URL_params(vars)}'>" \
                f" {(page_index + 1) if not label else label}" \
                f" </a>" \
                f"</li>"

    last_page_index = math.ceil(all_count / paginate) - 1
    html_str = ""
    if last_page_index > 0:
        page_number_btns = [create_link(current_page_index, active = True)]
        i = 0
        for i in range(current_page_index - 1, current_page_index - 5, - 1):
            if i < 0:
                break
            page_number_btns.insert(0, create_link(i))
        if i > 0:
```

```
            page_number_btns.insert(0, create_link(0, "<<"))
        for i in range(current_page_index + 1, current_page_index + 5):
            if i > last_page_index:
                break
            page_number_btns.append(create_link(i))
        if i < last_page_index:
            page_number_btns.append(create_link(last_page_index, ">>"))
        #dump html content
        html_str += " ".join(page_number_btns)
    return html_str


async def grid(
        db,
        request: Request,
        response: Response,
        query,
        fields = None,
        order_by = None,
        paginate = 20,
        row_render = default_row_render,
        header_render = default_header_render,
        foot_render = default_foot_render
):
    """
生成分页的表格,其原理是获取页面中的页码参数,以便查询数据库
并根据页码生成对应底部选择器
    :param db:
    :param request:
    :param response:
    :param query: 该表格对应的查询语句
    :param fields: 要呈现的字段,默认为全部字段
    :param order_by: 排序方式
    :param paginate: 每页数据条数
    :param row_render: 行渲染器,用于自定义行内容
    :param header_render: 表头渲染器,用于自定义表头
    :param foot_render: 底部选择器渲染器,用于自定义选择器
    :return:
    """
    page_index = int(request.get_query_var(b"page_index", b"0"))

    #获取符合条件的所有数据的条数,用于计算分页
    all_count = await db(query).count()

    #如果外部指定了字段,则呈现指定的字段数据
    if fields:
        db_rows = await db(query).select(
            *fields,
            limitby = (paginate * page_index, (page_index + 1) * paginate),
            orderby = order_by
```

```
        )
    # 如果外部没有指定字段,则呈现所有字段数据
    else:
        db_rows = await db(query).select(
            limitby = (paginate * page_index, (page_index + 1) * paginate),
            orderby = order_by
        )

    table_body_rows_html_content = ""
    # 渲染表中的行
    for r in db_rows:
        row_render_result = row_render(
            db, request, response, r, db_rows.field_names
        )
        if isinstance(row_render_result, Awaitable):
            row_render_result = await row_render_result

        if isinstance(row_render_result, str):
            row_content = row_render_result
        elif isinstance(row_render_result, Bytes):
            row_content = row_render_result.decode("utf - 8")
        else:
            row_content = ''
        table_body_rows_html_content += row_content

    rendered_header_content = ''
    rendered_header_result = header_render(
        db, request, response, db_rows.field_names
    )
    if inspect.isawaitable(rendered_header_result):
        rendered_header_content = await rendered_header_result
    else:
        rendered_header_content = rendered_header_result
    if isinstance(rendered_header_content, str):
        table_header_html_content = rendered_header_content
    elif isinstance(rendered_header_content, Bytes):
        table_header_html_content = rendered_header_content.decode("utf - 8")
    else:
        table_header_html_content = ""

    table_html_content = \
        f"<div>" \
        f"  <div>" \
        f"    <table class = 'data - grid table'>" \
        f"      <thead>{table_header_html_content}</thead>" \
        f"      <tbody>{table_body_rows_html_content}</tbody>" \
        f"    </table>" \
        f"  </div>" \
        f"  <div class = 'page - numbers - container'>" \
        f"    <nav>" \
```

```
            f"        <ul class = 'pagination'>" \
            f"          {foot_render(db, request, response, page_index, paginate, all_count)}" \
            f"        </ul>" \
            f"      </nav>" \
            f"    </div>" \
            f"</div>"

        return table_html_content
```

其中的 URL. dict_to_URL_params 工具函数源码如下：

```
"""
第 9 章/HomeSharing/app/utils/URL.py
"""

import URLlib.parse


def dict_to_URL_params(d):
    """
将字典转换为 URL 参数对字符串
    :param d:
    :return:
    """
    return '&'.join(
        '%s=%s' % (k, URLlib.parse.quote(str(v))) for k, v in d.items()
    )
```

之后使用该组件渲染表格可以省去大量重复的工作，对 memberships 函数的源码进行修改，修改后的代码如下：

```
"""
第 9 章/HomeSharing/app/controllers/admin.py
"""
from ..db.action_with_db import ActionWithDb
from ..db import auth
from ..db import data_grid


class memberships(ActionWithDb):

    @auth.require_membership("admin")
    async def execute(self, req, res):
        db = self.db
        grid = await data_grid.grid(
            db, req, res,
            db.auth_membership.id > 0,
            order_by = ~ db.auth_membership.id
        )
```

```
        await res.render(
            "admin/memberships.html", title = "关系表",
            grid = grid
        )
```

对应的模板文件源码修改后的代码如下：

```
<!-- 第 9 章/HomeSharing/app/views/admin/index.html -->
{ % extends "admin_layout.html" % }
{ % block body % }
<div style = "margin - top:1rem;">
    {{ grid }}
</div>
{ % endblock % }
```

对比可见，该组件的用法极为简单，在实际开发工作中，节省的工作量非常可观。

9.9 组管理

在组管理面板中，需要实现的功能有以下几个：

（1）列出所有组。

（2）编辑组介绍。

（3）添加组。

首先列出所有组，业务逻辑代码如下：

```
"""
第 9 章/HomeSharing/app/controllers/admin.py
"""
from cms4py.http import Request
from ..db import auth
from ..db import data_grid
from ..db.action_with_db import ActionWithDb


class groups(ActionWithDb):

    @auth.require_membership("admin")
    async def execute(self, req, res):
        async def header_render(db, req, res, fields):
            return await res.render_string(
                "admin/groups_header.html"
            )

        async def row_render(db, req, res, row, fields):
            return await res.render_string(
                "admin/groups_row.html",
```

```
                row = row
            )

        db = self.db
        grid = await data_grid.grid(
            db, req, res,
            db.auth_group.id > 0,
            order_by = ~db.auth_group.id,
            row_render = row_render,
            header_render = header_render
        )
        await res.render(
            "admin/groups.html",
            title = "用户组管理", grid = grid
        )
```

因为需要自定义表格行及表头，所以自定义了渲染器，对应表格行渲染器的模板文件代码如下：

```
<!-- 第9章/HomeSharing/app/views/admin/groups_row.html -->
<tr>
<td>{{ row.id }}</td>
<td>{{ row.role }}</td>
<td>{{ row.description }}</td>
<td style = "width:5rem">
<button data-group_id = "{{ row.id }}"
                data-group_description = "{{ row.description }}"
                class = "btn btn-primary btn-sm btn-edit-group">
编辑
</button>
</td>
</tr>
```

对应表头渲染器的模板文件代码如下：

```
<!-- 第9章/HomeSharing/app/views/admin/groups_header.html -->
<tr>
<th> id</th>
<th> role</th>
<th>说明</th>
<th></th>
</tr>
```

组表格主框架模板代码如下：

```
<!-- 第9章/HomeSharing/app/views/admin/groups.html -->
{% extends "admin_layout.html" %}
{% block body %}
```

```
<div style="margin-top:1rem;" class="card">
<div class="card card-header" style="position: relative;">
<button class="btn btn-link btn-add-group">
            +
</button>
用户组
</div>
    {{ grid }}
</div>

<script src="/static_files/dialogs.js"></script>
<script src="/static_files/admin_groups.js"></script>

<style>
    table, ul {
        margin: 0 !important;
    }

    .btn-add-group {
        position: absolute;
        top: 0;
        right: 0.5rem;
        font-size: 16pt;
        text-decoration: none !important;
        font-weight: bold;
    }
</style>
{% endblock %}
```

此表格渲染之后的效果如图 9-16 所示。

图 9-16 组管理界面渲染效果

为了实现"添加"与"编辑"功能，还需要再写一个名为 edit_group 的 Action 进行支持，前端与此 Action 通信的方式为 AJAX，edit_group 的源码如下：

```
"""
第 9 章/HomeSharing/app/controllers/admin.py
"""

from cms4py.http import Request
from ..db import auth
from ..db import data_grid
from ..db.action_with_db import ActionWithDb


class edit_group(ActionWithDb):

    async def execute(self, req: Request, res):
        db = self.db
        group_id = req.get_var_as_str(b"group_id")
        group_role = req.get_var_as_str(b"role")
        group_description = req.get_var_as_str(b"description")

        if not await auth.has_membership(db, req, res, 'admin'):
            await res.end(b"AccessDenied")
            return

        if not group_id:
            # 如果没有指定 id,则创建新组
            await db.auth_group.insert(
                role = group_role,
                description = group_description
            )
        else:
            # 如果已指定 id,则更新指定的组
            await db(db.auth_group.id == group_id).update(
                description = group_description
            )
        await res.end(b"ok")
```

前端与之通信的 JavaScript 代码如下：

```
//第 9 章/HomeSharing/app/static/static_files/admin_groups.js
(function () {

    /**
     * 呈现添加组对话框
     * @param { * } group_id
     * @param { * } description
     */
    function showAddGroupDialog() {
        var d = $ (`<div class = "modal fade" tabindex = " - 1" role = "dialog">
<div class = "modal - dialog modal - lg" role = "document">
<div class = "modal - content">
```

```
< div class = "modal - header">
< h5 class = "modal - title">编辑组介绍</h5 >
< button type = "button" class = "close" data - dismiss = "modal" aria - label = "Close">
< span aria - hidden = "true"> &times;</span >
</button >
</div >
< div class = "modal - body">
< table class = "table table - borderless">
< tbody >
< tr >
< td >组名</td >
< td >< input name = "role" class = "form - control"></td >
</tr >
< tr >
< td >说明</td >
< td >
< textarea rows = "5" name = "description"
                    class = "form - control"></textarea >
</td >
</tr >
</tbody >
</table >
</div >
< div class = "modal - footer">
< button type = "button" class = "btn btn - secondary" data - dismiss = "modal">关闭</button >
< button type = "button" class = "btn btn - primary btn - save">保存</button >
</div >
</div >
</div >
</div >`).modal({
            keyboard: true,
            backdrop: true,
        }).appendTo(document.body).on("hide.bs.modal", function (e) {
            $ (this).remove();
        });

        let descriptionInput = d.find("textarea[name = 'description']");
        let roleInput = d.find("input[name = 'role']");

        function addListeners() {
            d.find(".btn - save").click(function () {
                let description = descriptionInput.val();
                let role = roleInput.val();

                if (!role) {
                    dialogs.showMessageDialog("请填写组名");
                    return;
                }
```

```
                    $ .post(
                        "/admin/edit_group",
                        { role: role, description: description }
                    ).done(data => {
                        if (data == "ok") {
                            location.reload();
                        } else {
                            dialogs.showMessageDialog("编辑失败", "提示");
                        }
                    }).fail(e => {
                        dialogs.showMessageDialog("连接服务器失败", "提示");
                    });
                });
            }

            function init() {
                addListeners();
            }

            init();
            return d;
        }

        /**
         * 呈现编辑组信息对话框
         * @param { * } group_id
         * @param { * } description
         */
        function showEditGroupDialog(group_id, description) {
            var d = $ (`<div class="modal fade" tabindex="-1" role="dialog">
<div class="modal-dialog modal-lg" role="document">
<div class="modal-content">
<div class="modal-header">
<h5 class="modal-title">编辑组介绍</h5>
<button type="button" class="close" data-dismiss="modal" aria-label="Close">
<span aria-hidden="true">&times;</span>
</button>
</div>
<div class="modal-body">
<textarea rows="5" name="description"
              class="form-control">${description}</textarea>
</div>
<div class="modal-footer">
<button type="button" class="btn btn-secondary" data-dismiss="modal">关闭</button>
<button type="button" class="btn btn-primary btn-save">保存</button>
</div>
</div>
</div>
</div>`).modal({
```

```
            keyboard: true,
            backdrop: true,
        }).appendTo(document.body).on("hide.bs.modal", function (e) {
            $ (this).remove();
        });

        let contentInput = d.find("textarea[name = 'description']");


        function addListeners() {
            d.find(".btn - save").click(function () {
                $ .post(
                    "/admin/edit_group",
                    { group_id: group_id, description: contentInput.val() }
                ).done(data => {
                    if (data == "ok") {
                        location.reload();
                    } else {
                        dialogs.showMessageDialog("编辑失败", "提示");
                    }
                }).fail(e => {
                    dialogs.showMessageDialog("连接服务器失败", "提示");
                });
            });
        }

        function init() {
            addListeners();
        }

        init();
        return d;
    }

    function setStyles() {
        $ (".table").addClass("table - bordered");
    }

    function addListeners() {
        $ (".btn - add - group").click(showAddGroupDialog);
        $ (".btn - edit - group").click(e => {
            showEditGroupDialog(
                $ (e.target).data("group_id"),
                $ (e.target).data("group_description")
            );
        });
    }

    function main() {
        setStyles();
```

```
        addListeners();
    }

    main();
})();
```

该程序依赖的 dialogs.js 源码如下：

```
//第 9 章/HomeSharing/app/static/static_files/dialogs.js
(function () {
    window.dialogs = window.dialogs || {};

    /**
     * 呈现正在加载中的对话框，该功能运行需要依赖 Bootstrap
     * @param { * } msg
     */
    dialogs.showLoading = function (msg) {
        return $ (`<div class = "modal" tabindex = " - 1" role = "dialog">
<div class = "modal - dialog" role = "document">
<div class = "modal - content">
<div class = "modal - header">
<div class = "spinner - border" role = "status">
<span class = "sr - only"> Loading...</span>
</div>
</div>
<div class = "modal - body">
<p> $ {msg}</p>
</div>
</div>
</div>
</div>`).modal({
                keyboard: false,
                backdrop: "static",
            }).appendTo(document.body).on("hide.bs.modal", function (e) {
                $ (this).remove();
            });
    };


    /**
     * 呈现一个消息对话框，该功能运行需要依赖 Bootstrap
     */
    dialogs.showMessageDialog = function (msg, title = "", closeCallback, closeBtnClass =
"btn - danger") {
        return $ (`<div class = "modal fade" tabindex = " - 1" role = "dialog">
<div class = "modal - dialog" role = "document">
<div class = "modal - content">
<div class = "modal - header">
<h5 class = "modal - title"> $ {title}</h5>
```

```
<button type="button" class="close" data-dismiss="modal" aria-label="Close">
<span aria-hidden="true">&times;</span>
</button>
</div>
<div class="modal-body">
        ${msg}
</div>
<div class="modal-footer">
<button type="button" class="btn ${closeBtnClass}" data-dismiss="modal">关闭</button>
</div>
</div>
</div>
</div>`).modal({
            keyboard: true,
            backdrop: true,
        }).appendTo(document.body).on("hide.bs.modal", function (e) {
            $(this).remove();
            if (closeCallback) {
                closeCallback();
            }
        });
    };
})();
```

前端页面对组进行操作时，基于对话框和AJAX实现，因此页面不跳转，用户体验更好，效果如图9-17所示。

图9-17　添加组界面效果

9.10 用户管理

与用户相关的信息一般只能由用户修改，所以与用户管理相关的功能对于管理员来说最重要的是权限管理，本节将实现用户列表及管理用户权限功能。

为 auth.py 增加 add_membership 和 remove_membership 两个函数便于后续使用，其代码如下：

```
"""
第 9 章/HomeSharing/app/db/auth.py
"""
#此处省略部分代码

async def add_membership(db, request, response, user_id, role):
    """
添加用户关系
    :param db:
    :param request:
    :param response:
    :param user_id: 指定的用户 id
    :param role: 指定的权限组名
    :return:
    """
    group = (await db(db.auth_group.role == role).select()).first()
    if group:
        await db.auth_membership.insert(user_id = user_id, group_id = group.id)


async def remove_membership(db, request, response, user_id, role):
    """
移除用户关系
    :param db:
    :param request:
    :param response:
    :param user_id: 用户 id
    :param role: 权限组名
    :return:
    """
    group = (await db(db.auth_group.role == role).select()).first()
    if group:
        await db(
            (db.auth_membership.user_id == user_id) &
            (db.auth_membership.group_id == group.id)
        ).delete()
```

接下来实现用户列表，创建一个名为 users 的 Action，其代码如下：

```
"""
第 9 章/HomeSharing/app/controllers/admin.py
"""
from cms4py.http import Request
```

```python
from ..db import auth
from ..db import data_grid
from ..db.action_with_db import ActionWithDb


class users(ActionWithDb):

    @auth.require_membership("admin")
    async def execute(self, req, res):
        async def header_render(db, req, res, fields):
            return await res.render_string(
                "admin/users_header.html"
            )

        async def row_render(db, req, res, row, fields):
            return await res.render_string(
                "admin/users_row.html",
                row = row
            )

        db = self.db
        grid = await data_grid.grid(
            db, req, res,
            db.auth_user.id > 0,
            order_by = ~db.auth_user.id,
            row_render = row_render,
            header_render = header_render
        )
        await res.render(
            "admin/users.html",
            title = "所有用户", grid = grid
        )
```

模板 admin/users.html 的代码如下：

```html
<!-- 第 9 章/HomeSharing/app/views/admin/users.html -->
{% extends "admin_layout.html" %}
{% block body %}
<div style = "margin-top:1rem;" class = "card">
<div class = "card card-header" style = "position: relative;">
所有用户
</div>
    {{ grid }}
</div>

<style>
    table, ul {
        margin: 0 !important;
    }
</style>
{% endblock %}
```

模板 admin/users_header.html 的源码如下：

```
<!-- 第 9 章/HomeSharing/app/views/admin/users_header.html -->
<tr>
<th>id</th>
<th>用户名</th>
<th>邮箱</th>
<th>手机号</th>
<th></th>
</tr>
```

模板 admin/users_row.html 的源码如下：

```
<!-- 第 9 章/HomeSharing/app/views/admin/users_row.html -->
<tr>
<td>{{ row.id }}</td>
<td>{{ row.user_name }}</td>
<td>{{ row.user_email }}</td>
<td>{{ row.user_phone }}</td>
<td style="width:7rem">
<a href="/admin/manage_permissions?userid={{ row.id }}"
            class="btn btn-primary btn-sm btn-edit-group">
权限管理
</a>
</td>
</tr>
```

用户列表渲染效果如图 9-18 所示。

图 9-18 用户列表渲染效果

接下来实现权限管理功能页面，manage_permissions 的源码如下：

```
"""
第 9 章/HomeSharing/app/controllers/admin.py
"""
from cms4py.http import Request
from ..db import auth
from ..db import data_grid
from ..db.action_with_db import ActionWithDb


class manage_permissions(ActionWithDb):

    @auth.require_membership('admin')
    async def execute(self, request, response):
        user = await self.db.auth_user.by_id(
            request.get_var_as_str(b"userid")
        )
        if user:
            groups = await self.db(self.db.auth_group.id > 0).select()
            for g in groups:
                g.row_data['has_membership'] = await auth.has_membership(
                    self.db, request, response, g.role,
                    user.id
                )
            await response.render(
                "admin/manage_permissions.html",
                the_user = user,
                groups = groups,
                title = '权限管理'
            )
        else:
            await response.end(b"User not found.")
```

对应的 admin/manage_permissions.html 模板源码如下：

```
<!-- 第 9 章/HomeSharing/app/views/admin/manage_permissions.html -->
{% extends "admin_layout.html" %}
{% block body %}
<style>
    .table td {
        padding: 0;
    }
</style>
<div style = "margin-top:1rem;">
<div class = "card">
<div class = "card-header">
编辑用户权限
</div>
```

```
<div class="card-body">
<div style="display: flex;flex-direction: row;">
<table class="table table-borderless" style="width: 360px;">
<tbody>
<tr>
<td class="font-weight-bold" style="width: 80px;">
                                    id
</td>
<td>{{ the_user["id"] }}</td>
</tr>
<tr>
<td class="font-weight-bold" style="width: 80px;">
用户名
</td>
<td>{{ the_user["user_name"] }}</td>
</tr>
<tr>
<td class="font-weight-bold">邮箱</td>
<td>{{ the_user["user_email"] }}</td>
</tr>
<tr>
<td class="font-weight-bold">用机号</td>
<td>{{ the_user["user_phone"] }}</td>
</tr>
</tbody>
</table>
<div style="flex:1;border-radius: 10px" class="bg-light">
<table class="table table-borderless">
<thead>
<tr>
<th>id</th>
<th>role</th>
<th>描述</th>
<th>操作</th>
</tr>
</thead>
<tbody>
                                {% for g in groups %}
<tr>
<td>{{ g.id }}</td>
<td>{{ g.role }}</td>
<td>{{ g.description }}</td>
<td>
                                        {% if g.has_membership %}
<a href="/admin/edit_membership?role={{ g.role }}&userid={{ the_user['id'] }}&op=remove
_membership"
                                            class="btn btn-danger btn-sm"
                                            style="margin: 2px;">退出该组</a>
                                        {% else %}
```

```
<a href="/admin/edit_membership?role={{ g.role }}&userid={{ the_user['id'] }}&op=add_membership"
                                        class="btn btn-success btn-sm"
style="margin: 2px;">加入该组</a>
                                      {% endif %}
</td>
</tr>
                               {% endfor %}
</tbody>
</table>
</div>
</div>
</div>
</div>
</div>
{% endblock %}
```

权限管理页面渲染效果如图 9-19 所示。

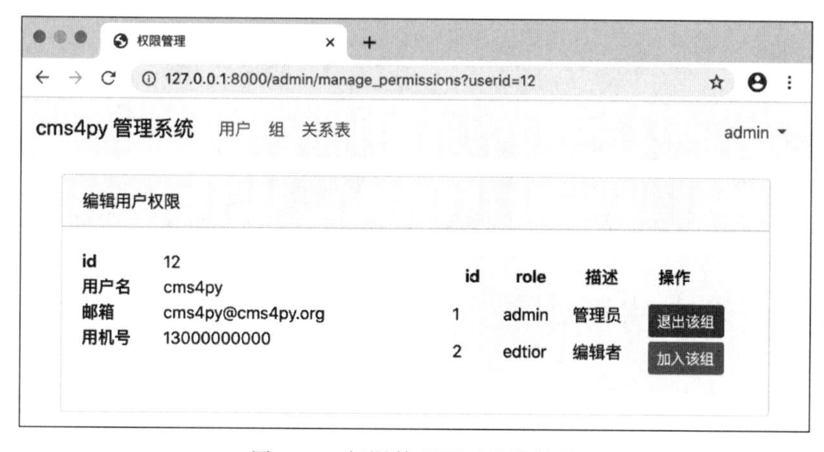

图 9-19　权限管理页面渲染效果

编辑用户关系页面/admin/edit_membership 的代码如下：

```
"""
第 9 章/HomeSharing/app/controllers/admin.py
"""
from cms4py.http import Request
from ..db import auth
from ..db import data_grid
from ..db.action_with_db import ActionWithDb

class edit_membership(ActionWithDb):

    @auth.require_membership('admin')
    async def execute(self, request, response):
        operation = request.get_var_as_str(b"op")
```

```
        role = request.get_var_as_str(b"role")
        userid = request.get_var_as_str(b"userid")
        current_user = await request.get_session("current_user")
        if operation == 'add_membership':
            await auth.add_membership(
                self.db, request, response, userid, role
            )
            await response.redirect(
                f"/admin/manage_permissions?userid={userid}"
            )
        elif operation == "remove_membership":
            if userid != str(current_user['id']) or role != 'admin':
                await auth.remove_membership(
                    self.db, request, response, userid, role
                )
                await response.redirect(
                    f"/admin/manage_permissions?userid={userid}"
                )
            else:
                await response.end(
                    b"You can't remove the admin membership of yourself"
                )
        else:
            await response.end(b"Unsupported action")
```

该功能完成后管理员可在权限管理页面自由配置用户的权限组。

9.11 实现发布房源功能

在发布房源功能中，需要一个富文本编辑器，这里使用 CKEditor 5，需要先通过地址 https://ckeditor.com/ckeditor-5/download/下载。下载后将其解压到静态文件目录中，如图 9-20 所示。

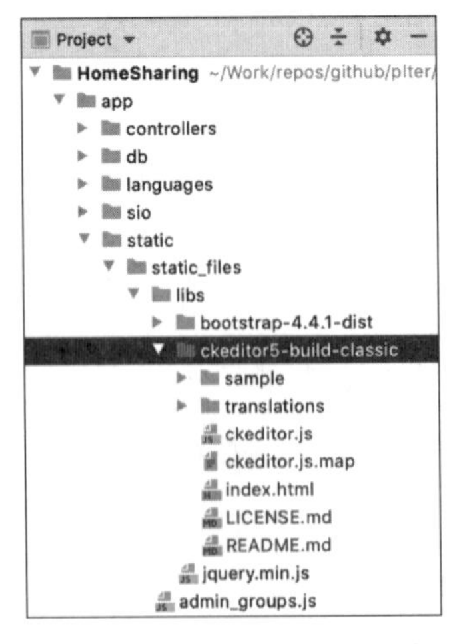

图 9-20 CKEditor 5 解压到静态文件目录中

发布房源的 Action(pub_house),代码如下:

```
"""
第 9 章/HomeSharing/app/controllers/user.py
"""

from app.db.action_with_db import ActionWithDb
from cms4py.http import Request, Response
import datetime, re
from pyMySQL.err import IntegrityError
from app.db import auth


class pub_house(ActionWithDb):
    @auth.require_login()
    async def execute(self, req: Request, res):
        if req.method == "GET":
            await res.render("user/pub_house.html", title="发布房源")
        elif req.method == 'POST':
            user = await req.get_session("current_user")
            title = req.get_var_as_str(b"house_res_title")
            content = req.get_var_as_str(b'house_res_content')
            await self.db.house_res.insert(
                res_title=title,
                res_content=content,
                pub_time=datetime.datetime.now(),
                owner_id=user['id']
            )
            await res.redirect(f"/house/by_id/{self.db.lastrowid}")
            pass
        else:
            await res.end(b"Method not supported")
```

其对应的模板文件 user/pub_house.html 源码如下:

```
<!-- 第 9 章/HomeSharing/app/views/user/pub_house.html -->
{% extends "layout.html" %}
{% block body %}
<script src="/static_files/libs/ckeditor5-build-classic/ckeditor.js"></script>
<div style="margin-top:1rem;">
<form enctype="multipart/form-data" method="POST">
<div style="margin-bottom: 0.5rem;">
<input placeholder="标题" required name="house_res_title" class="form-control">
</div>
<textarea name="house_res_content" placeholder="此处编写详细信息"></textarea>
<div style="margin-top: 0.5rem;">
<input type="submit" class="btn btn-primary" value="发布">
</div>
</form>
```

```
</div>

<script>
    ClassicEditor.create(
        document.querySelector('textarea[name="house_res_content"]'),
        {
            ckfinder: {
            uploadURL: `/ckeditor/file_upload_URL?command=QuickUpload&type=
Files&responseType=json`
            }
        }
    );
</script>
{% endblock %}
```

该页面渲染效果如图9-21所示。

图9-21　发布房源页面渲染效果

通过CKEditor 5将文本输入框渲染成一个富文本编辑器，用户可插入带格式的文字、图片、链接、表格等内容。在实际项目中，富文本编辑器是非常重要的组件，可适用于很多场景。

在该编辑器中，使用/ckeditor/file_upload_URL处理文件上传请求，接下来实现该Action，其代码如下：

```
"""
第9章/HomeSharing/app/controllers/ckeditor.py
"""
import json

from app.db.action_with_db import ActionWithDb
from cms4py.utils import aiofile
from app.db import auth
import os, config, datetime


class file_upload_URL(ActionWithDb):
```

```
    async def execute(self, req, res):
        user = await req.get_session("current_user")
        if not user:
            #如果用户没有登录,则提示错误
            await res.end(
                json.dumps(
                    dict(
                        error = "User not logged",
                        uploaded = False
                    )
                ).encode("utf-8")
            )
            return

        f = req.get_body_var(b"upload")
        current_datetime = datetime.datetime.now()
        #file_path 为文件路径,用于作为文件存储在服务器的相对路径
        file_path = "u{}d{}{}".format(
            user["id"],
            current_datetime.strftime("%Y%m%d%H%M%S"),
            current_datetime.microsecond
        )
        file_uri = "{}/{}".format(
            config.UPLOAD_FILE_BASE_URI,
            file_path
        )
        fw = await aiofile.open_async(
            os.path.join(
                config.UPLOADS_ROOT,
                file_path
            ),
            "wb"
        )
        await fw.write(f['content'])
        await fw.close()
        await self.db.photo.insert(
            photo_name = f['filename'],
            photo_path = file_path,
            photo_uri = file_uri,
            creator = user['id'],
            creation_time = current_datetime
        )
        await res.end(
            json.dumps(
                dict(
                    error = None,
                    uploaded = True,
                    URL = file_uri
                )
```

```
            ).encode("utf-8")
        )
```

该代码使用 config 配置文件的字段，必须添加该字段，因此需要在 config.py 文件中添加如下代码：

```
"""
第 9 章/cms4py_first_generation/config.py
"""

import os
#此处省略部分代码


#上传文件的根路径,用于浏览器访问
UPLOAD_FILE_BASE_URI = '/static_files/uploads'
#上传文件目录
UPLOADS_ROOT = os.path.join(STATIC_FILES_ROOT, "static_files", "uploads")
```

所有的上传文件将被自动存放在 app/static/static_files/uploads 目录，如图 9-22 所示。

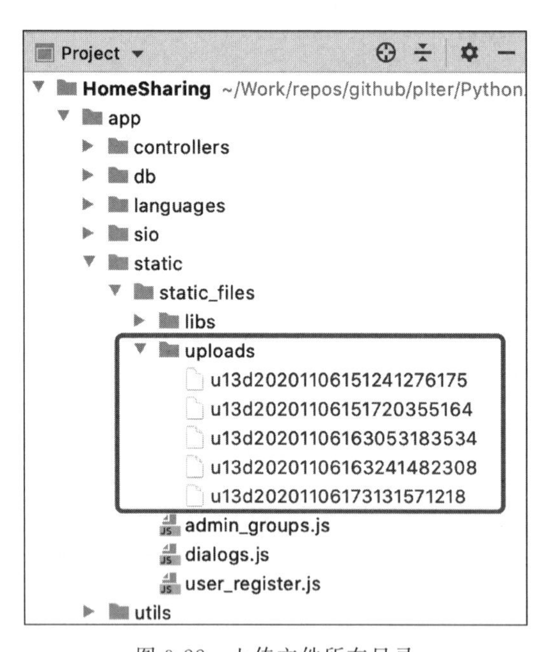

图 9-22　上传文件所在目录

在处理上传文件的 Action 中，除了将文件写入硬盘之外，还将相关的信息写入了数据库。在使用数据库之前，需要先在 DAL 中定义表，代码如下：

```
"""
第 9 章/HomeSharing/app/db/table_photo.py
"""
```

```
from pydal import Field


def define_table(db):
    db.define_table(
        "photo",
        Field('photo_name'),
        Field('photo_uri'),
        Field('photo_path'),
        Field('creator', type = "reference auth_user"),
        Field('creation_time', type = "datetime"),
    )
```

如果要使该定义生效，则需要在 app/db/__init__.py 文件中调用，示例代码如下：

```
"""
第 9 章/cms4py_first_generation/app/db/__init__.py
"""

from . import table_photo


class Db:
    # 此处省略部分代码

    async def _async_init(self):
# 此处省略部分代码
        table_photo.define_table(self._async_pydal)
        pass
```

如此可实现与上传图片相关的功能。在上传成功一张图片后，将直接呈现于富文本编辑器中，效果如图 9-23 所示。

图 9-23　上传图片的效果

接下来实现房源数据保存功能，在 pub_house 这个 Action 中，把数据存放到 house_res 表中，此时需要先在 DAL 中定义该表，代码如下：

```
"""
第 9 章/HomeSharing/app/db/table_house_res.py
"""

from pydal import Field


def define_table(db):
    db.define_table(
        "house_res",
        Field('res_title'),
        Field('res_content'),
        Field('pub_time', type = 'datetime'),
        Field('owner_id', type = "reference auth_user")
    )
```

如果要使该定义生效，则需要在 app/db/__init__. py 文件中调用，示例代码如下：

```
"""
第 9 章/cms4py_first_generation/app/db/__init__.py
"""
from . import table_house_res


class Db:
    # 此处省略部分代码

    async def _async_init(self):
        # 此处省略部分代码
        table_house_res.define_table(self._async_pydal)
        pass
```

接下来在主框架页 layout. html 中添加发布房源的链接，便于用户打开该页面，代码如下：

```
<!-- 第 9 章/HomeSharing/app/views/layout.html -->
...
<div class = "dropdown - menu dropdown - menu - right" aria - labelledby = "navbarDropdown">
    <a class = "dropdown - item" href = "/user/profile">我的信息</a>
    <a class = "dropdown - item" href = "/user/pub_house">发布房源</a>
    <div class = "dropdown - divider"></div>
    <a class = "dropdown - item" href = "/user/logout">登出</a>
</div>
...
```

渲染结果如图 9-24 所示。

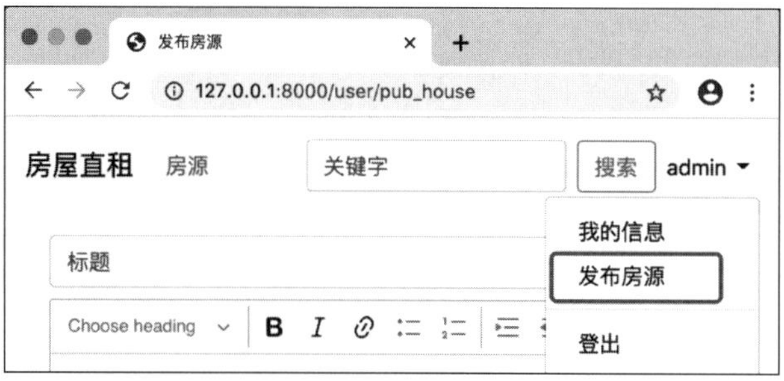

图 9-24 菜单渲染结果

在房源信息保存成功后，自动跳转到/house/by_id/{id}页面，在该页面中根据房源 id 直接查询出对应的房源并呈现出来，代码如下：

```python
"""
第 9 章/HomeSharing/app/controllers/house.py
"""
from app.db.action_with_db import ActionWithDb


class by_id(ActionWithDb):
    async def execute(self, req, res):
        house = await self.db.house_res.by_id(req.arg(0))
        owner = None
        if house:
            owner = await self.db.auth_user.by_id(house['owner_id'])
        await res.render(
            "house/by_id.html",
            title=house['res_title'] if house else '找不到房源',
            house=house,
            owner=owner
        )
```

对应的模板 house/by_id.html 源码如下：

```html
<!-- 第 9 章/HomeSharing/app/views/house/by_id.html -->
{% extends "layout.html" %}
{% block body %}
<div style="margin-top:1rem;">
    {% if house %}
<h1 style="font-size:14pt;font-weight:bold;">
        {{ house['res_title'] }}
</h1>
<div>
<span style="font-weight:bold;">发布者:</span>{{ owner['user_name'] }}
<span style="font-weight:bold;">手机号:</span>{{ owner['user_phone'] }}
<span style="font-weight:bold;">发布时间:</span>{{ house['pub_time'] }}
```

```
</div>
<div>
        {{ house['res_content'] }}
</div>
    {% else %}
<div>该房源不存在</div>
    {% endif %}
</div>
{% endblock %}
```

发布房源成功后效果如图 9-25 所示。

图 9-25　房源预览页面显示效果

9.12　房源列表

为了便于用户通过单击“房源”链接，需要为该链接设置真实的链接地址。对 layout.html 主框架中该链接的代码进行修改，修改后的代码如下：

```
<a class = "nav - link" href = "/house/all">房源</a>
```

单击该链接将跳转到/house/all 页面，该页面对应的 Action 业务逻辑代码如下：

```
"""
第 9 章/HomeSharing/app/controllers/house.py
"""
```

```
from app.db import data_grid
from app.db.action_with_db import ActionWithDb


class all(ActionWithDb):

    async def execute(self, req, res):
        async def row_render(db, req, res, row, fields):
            return await res.render_string(
                "house/all_row.html",
                row = row
            )

        db = self.db
        grid = await data_grid.grid(
            db, req, res,
            #多表查询
            (db.house_res.id > 0) &
            (db.house_res.owner_id == db.auth_user.id),
            fields = [
                #只显示指定的字段
                db.house_res.id,
                db.house_res.res_title,
                db.house_res.pub_time,
                db.auth_user.user_name,
                db.auth_user.user_phone,
            ],
            order_by = ~db.house_res.id,
            row_render = row_render,
            header_render = lambda *args: ""
        )
        await res.render(
            "house/all.html",
            title = "房源", grid = grid
        )
        pass
```

模板 house/all.html 文件的源码如下：

```
{% extends "layout.html" %}
{% block body %}
<div style = "margin-top:1rem;">
    {{ grid }}
</div>
{% endblock %}
```

模板 house/all_row.html 文件的源码如下：

```
<tr>
<td>
```

```
<a href="/house/by_id/{{ row['id'] }}"class="text-success">
            {{ row['res_title'] }}
</a>
<div>
<span style="font-weight:bold;">发布者:</span>{{ row['user_name'] }}
<span style="font-weight:bold;">发布时间:</span>{{ row['pub_time'] }}
</div>
</td>
</tr>
```

所有房源列表页面渲染效果如图 9-26 所示。

图 9-26　房源列表渲染效果

用户可通过单击房源链接进入房源详情页面。至此，发布房源的功能已完全实现。

9.13　实现搜索房源功能

写一个 Action 并命名为 search，用于实现搜索功能，将其放在 house.py 文件中。因为搜索结果也是以列表形式呈现，所以可以复用房源列表模板。而搜索关键字可以模糊匹配标题和内容，所以源码如下：

```
"""
第 9 章/HomeSharing/app/controllers/house.py
"""
from app.db import data_grid
from app.db.action_with_db import ActionWithDb
from cms4py.http import Request


class search(ActionWithDb):
    async def execute(self, req: Request, res):
        keyword = req.get_var_as_str(b"k")

        async def row_render(db, req, res, row, fields):
            return await res.render_string(
                "house/all_row.html",
```

```
            row = row
        )

    db = self.db
    grid = await data_grid.grid(
        db, req, res,
        #根据指定的关键字查询数据库
        (
                (db.house_res.res_title.like(f"%{keyword}%")) |
                (db.house_res.res_content.like(f"%{keyword}%"))
        ) &
        (db.house_res.owner_id == db.auth_user.id),
        fields = [
            #只显示指定的字段
            db.house_res.id,
            db.house_res.res_title,
            db.house_res.pub_time,
            db.auth_user.user_name,
            db.auth_user.user_phone,
        ],
        order_by = ~db.house_res.id,
        row_render = row_render,
        header_render = lambda *args: ""
    )
    await res.render(
        "house/all.html",
        title = f"搜索结果 - {keyword}", grid = grid
    )
    pass
```

接下来修改 layout.html 模板文件以启用搜索表单功能，示例代码如下：

```
<!-- 第 9 章/HomeSharing/app/views/layout.html -->
...
<form class = "form-inline my-2 my-lg-0"
    action = "/house/search" method = "get">
    <input name = "k"
    class = "form-control mr-sm-2"
    type = "search" placeholder = "关键字"
    aria-label = "Search">
    <button class = "btn btn-outline-success my-2 my-sm-0" type = "submit">
    搜索
    </button>
</form>
...
```

渲染之后的搜索效果如图 9-27 所示。

图 9-27　搜索功能效果

9.14　实现房源评论功能

房源评论功能在房源详情页面继续开发，并且只有在用户处于登录状态下才能发表评论，非登录状态下将呈现登录按钮，如图 9-28 所示。

图 9-28　非登录状态下显示的评论框

原理是先判断用户是否处于非登录状态，如果用户处于非登录状态，则在评论框前面呈现一个背景半透明的 DIV，将登录按钮放在该 DIV 中，实现该界面的模板文件源码如下：

```
<!-- 第 9 章/HomeSharing/app/views/house/by_id.html -->
{ % extends "layout.html" % }
{ % block body % }
<div style = "margin - top:1rem;">
    { % if house % }
```

```
<h1 style="font-size:14pt;font-weight:bold;">
        {{ house['res_title'] }}
</h1>
<div>
<span style="font-weight:bold;">发布者:</span>{{ owner['user_name'] }}
<span style="font-weight:bold;">手机号:</span>{{ owner['user_phone'] }}
<span style="font-weight:bold;">发布时间:</span>{{ house['pub_time'] }}
</div>
<div style="margin-top:1rem;">
        {{ house['res_content'] }}
</div>
<script src="/static_files/libs/ckeditor5-build-classic/ckeditor.js"></script>
<div style="position:relative">
<form method="post" enctype="multipart/form-data">
<textarea name="comment"></textarea>
<input type="submit" class="btn btn-success" value="发表评论" style="margin-top:
0.5rem">
</form>
        {% if 'current_user' not in session or not session['current_user'] %}
<div class="editor-overlay">
<a href="/user/login" class="btn btn-success" style="margin-top:1rem;">
登录后发表评论
</a>
</div>
        {% endif %}
</div>
<div style="margin-top:0.5rem">
        {{ comments_grid }}
</div>
<script>
        ClassicEditor.create(
            document.querySelector('textarea[name="comment"]'),
            {
                toolbar:[
                    'heading', '|',
                    'bold', 'italic', 'indent', 'outdent',
                    'numberedList', 'bulletedList',
                    'blockQuote', '|',
                    'undo', 'redo'
                ]
            }
        );

</script>
<style>
        .editor-overlay{
            position:absolute;
            width:100%;height:100%;
            left:0;top:0;
            background-color:rgba(255,255,255,0.7);
```

```
            text - align:center;
        }

</style>
    {% else %}
<div>该房源不存在</div>
    {% endif %}
</div>
{% endblock %}
```

对应的业务逻辑处理代码如下：

```
"""
第 9 章/HomeSharing/app/controllers/house.py
"""
from app.db import data_grid
from app.db.action_with_db import ActionWithDb
from cms4py.http import Request
import datetime


class by_id(ActionWithDb):
    async def execute(self, req: Request, res):
        house_id = req.arg(0)
        house = await self.db.house_res.by_id(house_id)
        owner = None
        comments_grid = ""

        async def row_render(db, req, res, row, fields):
            return await res.render_string(
                "house/comment_row.html",
                row = row
            )

        if house:
            owner = await self.db.auth_user.by_id(house['owner_id'])

            if req.method == 'POST':
                # 如果是 POST 请求,则尝试读取评论数据
                comment = req.get_var_as_str(b"comment")
                if comment:
                    await self.db.house_res_comment.insert(
                        user_id = owner['id'],
                        house_res_id = house['id'],
                        comment_content = comment,
                        comment_time = datetime.datetime.now()
                    )
            db = self.db
            comments_grid = await data_grid.grid(
```

```
                db, req, res,
                (db.house_res_comment.house_res_id == house['id']) &
                (db.house_res_comment.user_id == db.auth_user.id),
                fields = [
                    # 只显示指定的字段
                    db.house_res_comment.id,
                    db.house_res_comment.comment_content,
                    db.house_res_comment.comment_time,
                    db.auth_user.user_name
                ],
                order_by = ~db.house_res_comment.id,
                row_render = row_render,
                header_render = lambda *args: ""
            )
        await res.render(
            "house/by_id.html",
            title = house['res_title'] if house else '找不到房源',
            house = house,
            owner = owner,
            comments_grid = comments_grid
        )
```

评论行的渲染采用了单独的模板文件 comment_row.html，其源码如下：

```
<!-- 第 9 章/HomeSharing/app/views/house/comment_row.html -->
<tr>
<td>
<div>
<span style = "font-weight:bold;">{{ row['user_name'] }} </span>
            {{ row['comment_time'] }}
</div>
<div>
            {{ row['comment_content'] }}
</div>
</td>
</tr>
```

评论内容需要保存到数据库中，这需要先在 DAL 中定义 house_res_comment 表，定义源码如下：

```
"""
第 9 章/HomeSharing/app/db/table_house_res_comment.py
"""

from pydal import Field


def define_table(db):
```

```
    db.define_table(
        "house_res_comment",
        Field('user_id', type = "reference auth_user"),
        Field('house_res_id', type = "reference house_res"),
        Field('comment_content'),
        Field('comment_time', type = 'datetime')
    )
```

若要启用该定义，需要在 app/db/ __init__. py 文件中调用，该文件完整源码如下：

```
"""
第 9 章/cms4py_first_generation/app/db/__init__.py
"""

from app.db.async_pydal.dal import AsyncDAL
from . import table_auth_group
from . import table_auth_membership
from . import table_auth_user
from . import table_photo
from . import table_house_res
from . import table_house_res_comment


class Db:
    __instance = None

    @staticmethod
    async def get_instance() -> "Db":
        if not Db.__instance:
            Db.__instance = Db()
            await Db.__instance._async_init()
        return Db.__instance

    async def _async_init(self):
        self._async_pydal = await AsyncDAL.create(
            "db", "root", "rootpw", "home_sharing"
        )

        table_auth_user.define_table(self._async_pydal)
        table_auth_group.define_table(self._async_pydal)
        table_auth_membership.define_table(self._async_pydal)
        table_photo.define_table(self._async_pydal)
        table_house_res.define_table(self._async_pydal)
        table_house_res_comment.define_table(self._async_pydal)
        pass

    @property
    def async_pydal(self):
        return self._async_pydal

    pass
```

评论列表渲染的效果如图 9-29 所示。

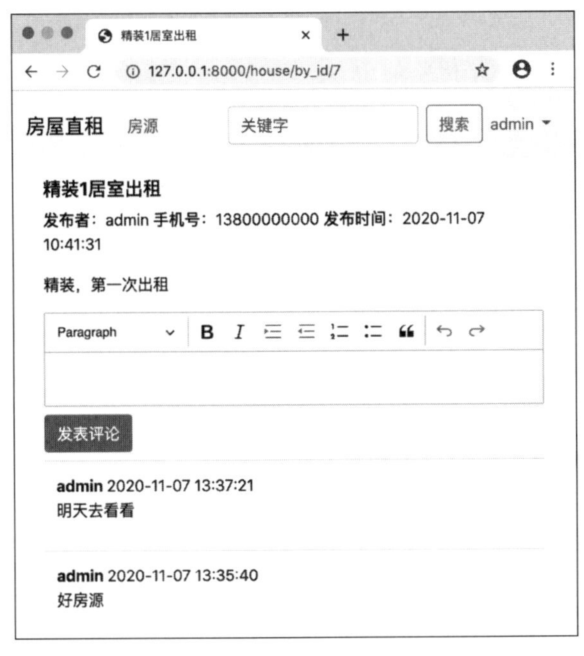

图 9-29 评论列表渲染效果

9.15 部署项目

由于采用了 Docker 容器化技术，项目部署和迁移工作都易于进行。这里笔者以一台网址为 192.168.8.179 的服务器（读者需要自建服务器或者购买云主机）为例进行讲解，服务器系统为 Linux，登录该服务器的用户名为 s。使用命令 ssh s@192.168.8.179 登录，如图 9-30 所示。

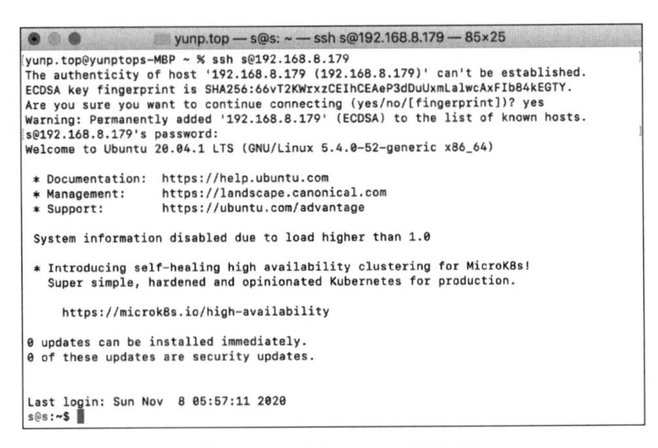

图 9-30 登录 Linux 服务器

Windows 计算机上可能没有内置 ssh 命令，可先安装 Git 终端（https://git-scm.com/downloads），Git 终端会自带一个 MSYS 环境，其中有 ssh 命令。

接下来备份数据库，在本机 Docker 数据库处于运行状态下执行命令 docker-compose exec db MySQLdump -uroot -prootpw --Databases home_sharing > home_sharing.sql 以备份数据库。

之后使用 FileZilla（下载地址 https://filezilla-project.org/download.php）通过 SFTP 的方式连接服务器。通过菜单 File → SiteManager（文件→站点管理）打开连接管理面板，如图 9-31 和图 9-32 所示。

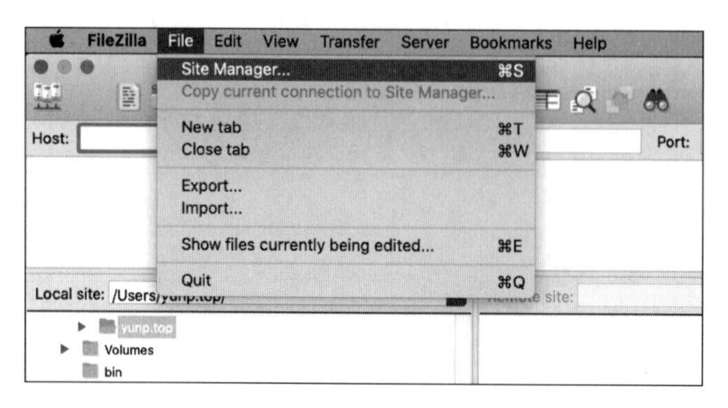

图 9-31 打开连接管理面板

图 9-32 连接管理面板

单击 New site（新站点）按钮以对连接进行配置，Protocol（协议）选择 SFTP，用户名及密码按实际情况配置，如图 9-33 所示。

连接成功后将项目目录 HomeSharing 映射到服务器端的~/HomeSharing 目录，并将 app、cms4py、docker、config.py、docker-compose.yml、home_sharing.sql 上传到服务器中，如图 9-34 所示。

使用 ssh 命令登录服务器，并切换目录到~/HomeSharing，执行命令 sudo docker-compose up -d 以启动服务器，如图 9-35 所示。

然后使用命令 sudo docker-compose exec -T db MySQL -uroot -prootpw < ./home_sharing.sql 将数据恢复到数据库中。

最后通过网址 http://192.168.8.179/house/all 访问该系统。

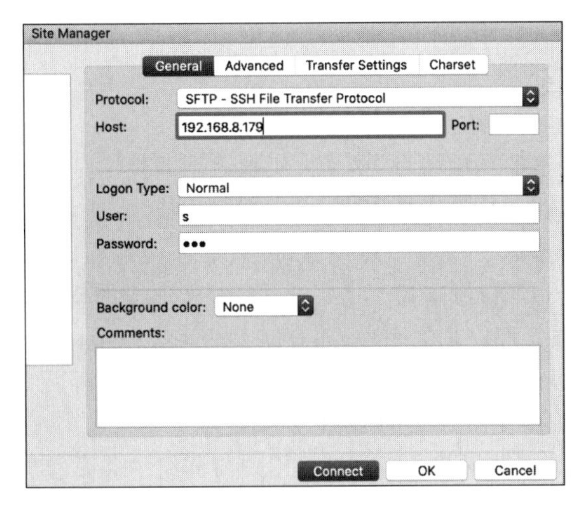

图 9-33　SFTP 连接配置

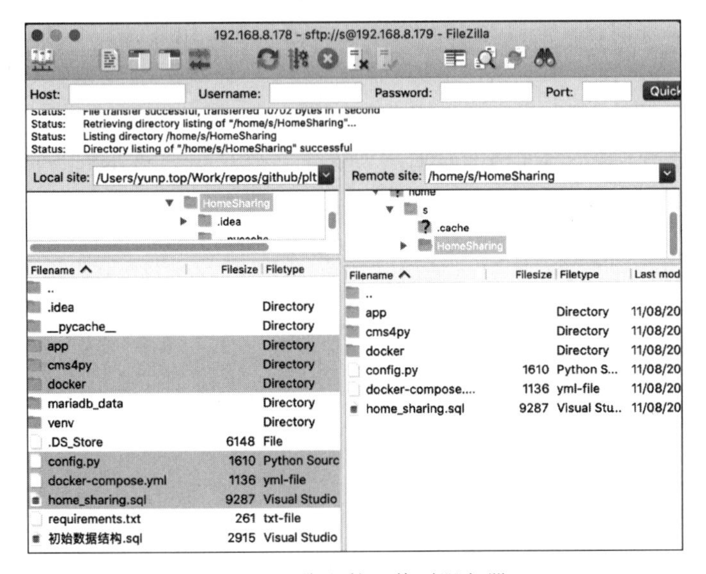

图 9-34　将文件上传到服务器

```
yunp.top — s@s: ~/HomeSharing — ssh s@192.168.8.179 — 86×25
Status: Downloaded newer image for mariadb:latest
Pulling adminer (adminer:)...
latest: Pulling from library/adminer
188c0c94c7c5: Pull complete
45f8bf6cfdbe: Pull complete
ce5be7974012: Pull complete
a99dd6507fe5: Pull complete
84ab7a98c2ef: Pull complete
07fa46c284ad: Pull complete
7ae9397c88ea: Pull complete
5e4748577090: Pull complete
36fcf1c47c04: Pull complete
a190b0137493: Pull complete
e33b1564d2df: Pull complete
2ec83214bdcb: Pull complete
9143e4447235: Pull complete
1a15263e4f11: Pull complete
1a3ca31264b7: Pull complete
Digest: sha256:94f2f8703432f6cf80b4db8f8e6209b68541d815fb86713f0126540c638a1c7f
Status: Downloaded newer image for adminer:latest
Creating homesharing_gateway_1 ... done
Creating homesharing_web_1     ... done
Creating homesharing_adminer_1 ... done
Creating homesharing_db_1      ... done
s@s:~/HomeSharing$
```

图 9-35　启动 Docker 环境

9.16 项目总结

这个项目是一个完整的基础项目，尽量去实现通用的技术，所以可被用于二次开发以便快速完成其他常规项目。运行环境采用 Docker 容器化技术，便于部署和迁移。

在设计架构时充分考虑了动静分离，所以可以很方便地集成 Apache 服务器。在选择 Apache 服务器镜像时，笔者特意选择了已集成了 PHP 环境的镜像，在实际项目中，这非常实用。PHP 也是一个强大而完整的平台，拥有庞大的生态体系，现今流行的各种 CMS 框架大多是由 PHP 语言开发的，例如大名鼎鼎的 Wordpress。这样除了可以利用 Apache 处理静态资源请求外，还能集成更多成熟的 PHP 框架技术。

在权限管理的设计上，做到了至简。虽然用户与群组的关系管理相对比较复杂，但开发人员在使用这些功能时仅需要写一个装饰器即可完成。

在数据库的使用上，笔者尽最大可能介绍了 pyDAL 的各种常规用法，包括根据 id 查询、模糊查询、多表查询、删除、插入、修改等，这都是非常实用并且易于拓展的技术。

在模板渲染上，全面介绍了条件渲染、循环渲染、模板继承等技术的运用。同时还实现了一个通用的分页组件，大大提高了开发效率。

在处理基本业务逻辑上，介绍的是通用的技术。运用得当可很容易地将该项目改成博客、论坛、门户网站、新闻站等。

附录A 名词解释

AIO,Asynchronous I/O,翻译为异步IO,在Python语言中的库名为asyncio。

BIO,Blocked I/O,阻塞型IO,是传统的IO模型。

WSGI(Web Server Gateway Interface),翻译为服务器网关接口,是为Python语言定义的Web服务器和Web应用程序或框架之间的一种简单而通用的接口。

ASGI(Asynchronous Server Gateway Interface),翻译为异步服务网关接口,是WSGI精神的继任者,在异步服务、框架和应用之间提供一个标准接口,同时兼容WSGI。

JIT(Just in Time),在运行时将部分代码编译成机器码执行,从而提高运行速度。

API(Application Programming Interface),应用开发编程接口。

前端与后端,程序员通常说的前端是指Web前端,是狭义的,但在本书中,前端与后端的说法是相对的,是广义的,在客户端与服务器(C-S)模式下,客户端是前端,服务器是后端,在浏览器与服务器(B-S)模式下,浏览器是前端,服务器是后端,在Apache代理Python应用服务器的架构模式下,Apache服务器相对Python服务器是前端,Python服务器相对Apache服务器是后端。

附录 B

开发环境约定

本书中所讲述的所有示例将使用下列软件或者平台进行开发。

操作系统要求：

- Windows 10 2004 或更高版本。
- Ubuntu 2004 或更高版本。
- Mac OS X 10.15 或更高版本。

Python 版本：

- Python 3.8 或更高版本。

集成开发环境：

- PyCharm Community 2020.1 或更高版本(读者也可以使用 PyCharm 专业版进行学习)。
- Visual Studio Code 1.49.2 或更高版本。

附录C 创建项目及依赖项安装

本书的所有示例中，如果有第三方依赖项，需要使用 pip 进行安装，下面演示在 PyCharm Community 环境中的操作方式。

在 PyCharm Community 的欢迎界面单击 Create New Project 按钮，如图 C-1 所示。

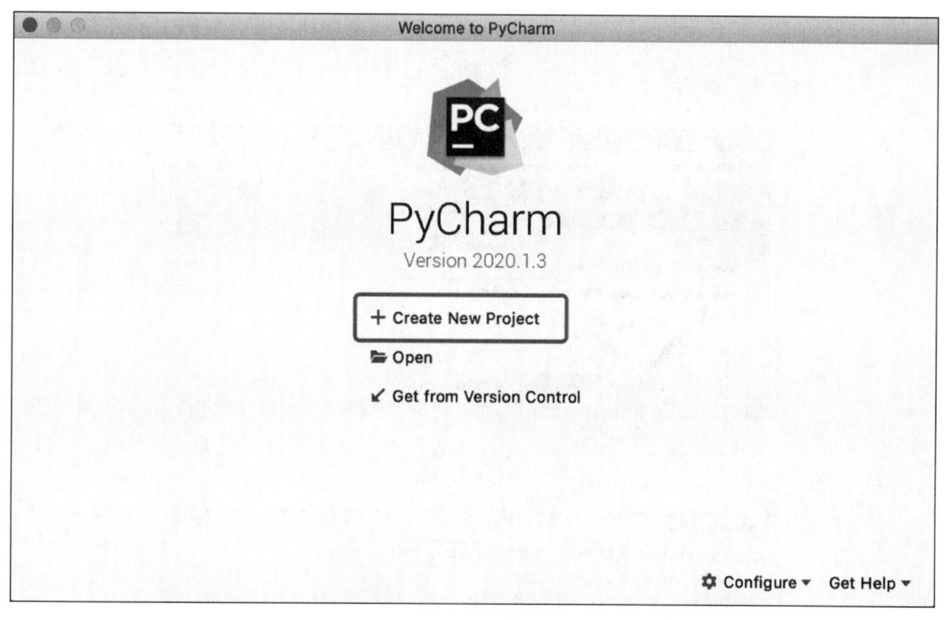

图　C-1

在创建项目界面将项目命名为 hello_world，并为该项目创建一个基于 Python 3.8 的虚拟环境，如图 C-2 所示。

通过这种方式创建的项目在编码界面中打开终端视图时将自动激活随项目一起创建的虚拟环境，如图 C-3 所示。

在该终端里使用 pip 安装的依赖项目将自动安装在与项目相关的虚拟环境中。图 C-4 所示为安装第三方依赖项 NumPy。

因为安装第三方依赖项需要联网，而 pip 的服务器在国外，所以速度比较慢，国内用户建议使用清华提供的 pip 镜像进行安装，以获取更快的下载及安装速度。使用清华 pip 镜像安装 NumPy 效果如图 C-5 所示。

New Project

Location: /Users/yunp/Work/gitroot/github/plter/PythonAIOProgramming/hello_world

Project Interpreter: New Virtualenv environment

New environment using Virtualenv

Location: /Users/yunp/Work/gitroot/github/plter/PythonAIOProgramming/hello_world/venv

Base interpreter: /usr/local/bin/python3.8

Inherit global site-packages

Make available to all projects

Existing interpreter

Interpreter: <No interpreter>

Cancel Create

图 C-2

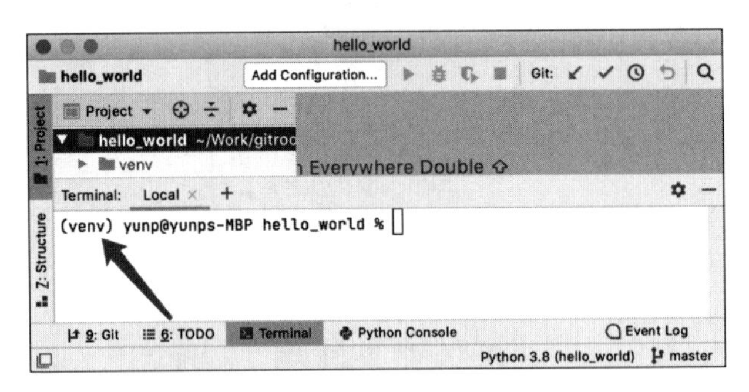

图 C-3

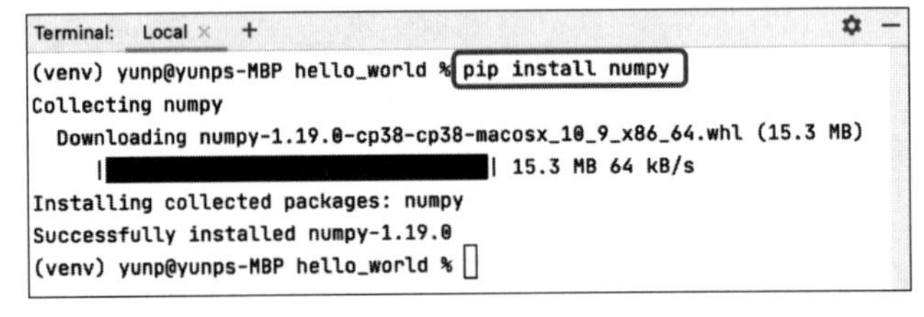

图 C-4

```
Terminal: Local
(venv) yunp@yunps-MBP hello_world % pip install -i https://pypi.tuna.tsinghua.edu.cn/simple numpy
Looking in indexes: https://pypi.tuna.tsinghua.edu.cn/simple
Collecting numpy
  Using cached https://pypi.tuna.tsinghua.edu.cn/packages/fe/89/92751677843f0a81a10d0befff7cb213d830956
6c14216a1c57f01b6fd39/numpy-1.19.0-cp38-cp38-macosx_10_9_x86_64.whl (15.3 MB)
Installing collected packages: numpy
Successfully installed numpy-1.19.0
(venv) yunp@yunps-MBP hello_world %
```

图 C-5

参 考 文 献

[1] Python 官方文档[EB/OL]. [2020-03-28]. https://docs.python.org.

[2] Mozilla. HTTP 说明文档[EB/OL]. [2020-04-10]. https://developer.mozilla.org/en-US/docs/Web/HTTP.

[3] Docker 官方文档[EB/OL]. [2020-04-01]. https://docs.docker.com.

[4] AIOHTTP 官方文档[EB/OL]. [2020-04-20]. https://docs.aiohttp.org/en/stable/index.html.

[5] Nodejs 官方文档[EB/OL]. [2020-04-20]. https://nodejs.org/en/docs.

[6] Nikolay Novik. aiomysql 官方文档[EB/OL]. [2020-05-01]. https://aiomysql.readthedocs.io.

[7] MariaDB 官方文档[EB/OL]. [2020-05-01]. https://mariadb.org.

[8] MySQL 官方文档[EB/OL]. [2020-05-01]. https://www.mysql.com.

[9] SQLAlchemy 官方文档[EB/OL]. [2020-05-06]. https://docs.sqlalchemy.org.

[10] ASGI 官方文档[EB/OL]. [2020-06-01]. https://asgi.readthedocs.io.

[11] Graham Dumpleton. mod_wsgi 官方文档[EB/OL]. [2020-06-02]. https://modwsgi.readthedocs.io.

[12] 清华镜像站 Debian 镜像使用文档[EB/OL]. [2020-06-02]. https://mirrors.tuna.tsinghua.edu.cn/help/debian.

[13] Uvicorn 官方文档[EB/OL]. [2020-06-10]. http://www.uvicorn.org.

[14] DaphneREADME[EB/OL]. [2020-06-10]. https://github.com/django/daphne.

[15] Django 官方文档[EB/OL]. [2020-06-10]. https://docs.djangoproject.com/en/3.1.

[16] Quart 官方文档[EB/OL]. [2020-06-12]. https://pgjones.gitlab.io/quart.

[17] Starlette 官方文档[EB/OL]. [2020-06-12]. https://www.starlette.io.

[18] Tornado 官方文档[EB/OL]. [2020-07-01]. https://www.tornadoweb.org.

[19] Socket.IO 官方文档[EB/OL]. [2020-07-20]. https://socket.io/docs.

[20] Miguel Grinberg. python-socketio 官方文档[EB/OL]. [2020-08-05]. https://python-socketio.readthedocs.io.

[21] Massimo Di Pierro. web2py 官方文档[EB/OL]. [2020-03-28]. http://web2py.com.

[22] Jinja2 官方文档[EB/OL]. [2020-04-20]. https://jinja.palletsprojects.com.

[23] CKEditor5 官方文档[EB/OL]. [2020-09-10]. https://ckeditor.com/docs/ckeditor5/latest.

图书资源支持

感谢您一直以来对清华版图书的支持和爱护。为了配合本书的使用，本书提供配套的资源，有需求的读者请扫描下方的“书圈”微信公众号二维码，在图书专区下载，也可以拨打电话或发送电子邮件咨询。

如果您在使用本书的过程中遇到了什么问题，或者有相关图书出版计划，也请您发邮件告诉我们，以便我们更好地为您服务。

我们的联系方式：

地　　址：北京市海淀区双清路学研大厦 A 座 714

邮　　编：100084

电　　话：010-83470236　010-83470237

客服邮箱：2301891038@qq.com

QQ：2301891038（请写明您的单位和姓名）

资源下载：关注公众号“书圈”下载配套资源。

书圈

获取最新书目

观看课程直播